Integrating Neuroimaging, Computational Neuroscience, and Artificial Intelligence

Unveil the next frontier in neurodegenerative disorder research with *Integrating Neuroimaging, Computational Neuroscience, and Artificial Intelligence*. This groundbreaking book goes beyond traditional approaches, utilizing the power of interdisciplinary integration to illuminate new pathways in diagnosis and treatment. From artificial intelligence (AI)-driven diagnostics to computational neuroscience models, this book showcases the forefront of innovation. Join us in exploring the future of neurodegenerative care, where collaboration and cutting-edge technology converge to redefine possibilities.

Key Features:

- Integrates diverse fields of research, from neuroimaging to computational neuroscience and AI.
- Emphasizes the translation of research findings into practical applications, ultimately benefitting patients and clinical practice.
- Reviews the implementation of AI and computational models in diagnostic settings.
- Elucidates the current state of translational neuroscience exploring potential areas for further research and collaboration, including personalized treatments and drug development.
- Contributions from an international group of experts from diverse disciplines.

Neuroscience of Neurodegenerative Disorders
Series Editor Wael Mohamed

Wael Mohamed Yousef Mohamed
BBCH (Egypt), MMSc (Egypt), MD/PhD (USA)
Dr. Mohamed earned his PhD from Penn State University (USA) and currently is an Associate Professor at IIUM Medical School in Malaysia. He serves as an editor for *Frontiers in Neurology* and PLOS-ONE Journals. His research is well-supported by extramural funding. He is the founder and chair of the AfrAbia-PD-Genomic Consortium.

About This Series
This book series aims to publish volumes focusing on both basic and clinical research in the field of Neurodegenerative Disorders with a focus on delineating the recent advances in various disorders including Parkinson's Disease, Alzheimer's Disease, ALS, Huntington's Disease, among other common Neurodegenerative disorders. The books will examine different disorders aspects including pathophysiology, diagnostic, treatment, and precision medicine. Volumes in the series will also highlight the recent advances in the management of Neurodegenerative disorders especially in OMICS and Genomic era and precision medicine. Volumes will be useful for neuroscientists, neurologists, psychiatrists, psychologists, neurosurgeons, and those interested in preventive medicine.

Integrating Neuroimaging, Computational Neuroscience, and Artificial Intelligence
Edited by Indranath Chatterjee and Nasrollah Moradikor

For more information about this book series, please visit: https://www.routledge.com/Neuroscience-of-Neurodegenerative-Disorder/book-series/NNDI

Integrating Neuroimaging, Computational Neuroscience, and Artificial Intelligence

Edited by
Indranath Chatterjee and Nasrollah Moradikor

CRC Press
Taylor & Francis Group
Boca Raton London New York

CRC Press is an imprint of the
Taylor & Francis Group, an **informa** business

Designed cover image: Shutterstock contributor korkeng

First edition published 2025
by CRC Press
2385 NW Executive Center Drive, Suite 320, Boca Raton FL 33431

and by CRC Press
4 Park Square, Milton Park, Abingdon, Oxon, OX14 4RN

CRC Press is an imprint of Taylor & Francis Group, LLC

ISBN: 9781032701899 (hbk)
ISBN: 9781032711096 (pbk)
ISBN: 9781032711102 (ebk)

DOI: 10.1201/9781032711102

Typeset in Times
by KnowledgeWorks Global Ltd.

Contents

Series Editor Preface

The human brain, a highly intricate organ, continues to be one of the most mysterious and demanding areas of study in the field of medical research. Neurodegenerative illnesses, including Alzheimer's disease, Parkinson's disease, Huntington's disease, and amyotrophic lateral sclerosis (ALS), are very debilitating and poorly understood conditions that have a significant impact on human health. Despite notable progress in the field of neurology, researchers and clinicians are still faced with the ongoing problem of finding effective treatments and, eventually, cures for these debilitating illnesses.

The purpose of this series, titled "Neuroscience of Neurodegenerative Disease," is to connect the divide between laboratory research and practical application, providing a thorough analysis of the present state of research, clinical methods, and future prospects in combating neurodegenerative disorders. Our objective is to provide a comprehensive examination of the complex processes that underlie these illnesses, the cutting-edge strategies being devised to comprehend and treat them, and the translational initiatives that are facilitating the application of scientific findings from the laboratory to clinical practice.

Translational neuroscience is focused on transforming fundamental scientific discoveries into tangible applications that improve human health and overall welfare. It is a multidisciplinary undertaking that needs the cooperation of neuroscientists, physicians, pharmacologists, geneticists, and several other experts. This series is specifically crafted to showcase the wide range of opinions and demonstrate the collaborative efforts that propel advancements in this industry.

Neuroimaging, computational neuroscience, and artificial intelligence (AI) are revolutionizing medical research. These exceptional technologies provide unmatched diagnosis and understanding of neurological illnesses. The book titled *Integrating Neuroimaging, Computational Neuroscience, and Artificial Intelligence* explores the potential of combining these approaches to revolutionize the field of brain disorder research and treatment. Neuroimaging enables us to see the intricate structure and dynamic functions of the brain. Magnetic resonance imaging (MRI), functional magnetic resonance imaging (fMRI), positron emission tomography (PET), and electroencephalography (EEG) are essential tools for neurological diagnosis and research. These technologies generate intricate and extensive datasets that need sophisticated analytical methods to extract insights. Computational neuroscience offers theoretical frameworks and mathematical models to understand neuroimaging data, while AI offers advanced machine learning and pattern recognition. They unveil hitherto undiscovered biomarkers, forecasting patterns, and links between the brain and behavior.

This book offers advanced research and practical applications to enhance diagnosis accuracy, predict the evolution of illnesses, and customize treatment choices. These sections demonstrate the impact of interdisciplinary cooperation on patient care and outcomes, emphasizing that it goes beyond the use of technology. This book explores not just scientific and technical advancements but also their ethical and societal implications. The use of AI and computational methods in neuroimaging gives rise to problems around the privacy of data, bias in algorithms, and the need for transparent and equitable healthcare. These aspects are crucial for the future of medical diagnosis and therapy. This book features contributions from leading experts in this fascinating field. Their research uncovers the potential and constraints of neuroscience and medical diagnostics, shedding light on their future implications.

I thus present this book with enthusiasm and reverence. *Integrating Neuroimaging, Computational Neuroscience, and Artificial Intelligence* serves as a testament to our advancements and a beacon illuminating the path for the future of medical research. May it catalyze creativity and foster collaboration to enhance the well-being of all individuals.

As the editor of this series, I express my gratitude to the contributors whose skill and commitment have made this endeavor possible. Their research not only enhances our comprehension of neurodegenerative illnesses but also exemplifies the essence of translational science—converting knowledge into concrete advantages for society. I would like to express my appreciation to the readers who actively participate in this series, as they contribute to the continuous effort of understanding the complexities of the brain and enhancing the well-being of those impacted by neurodegenerative disorders.

Through this series, our aim is to shed light on the way forward, bringing us nearer to a future when neurodegenerative conditions are comprehended, manageable, and eventually, able to be cured.

Sincerely,

Wael Mohamed, Series Editor
Book Series - Neuroscience of Neurodegenerative Disease

Preface

In an increasingly aging world, the burden of neurodegenerative disorders has become a paramount concern in modern healthcare. Conditions such as Alzheimer's disease, Parkinson's disease, and other neurodegenerative ailments affect millions of individuals worldwide, necessitating innovative approaches to understand, diagnose, and potentially treat these complex disorders. This book, *Integrating Neuroimaging, Computational Neuroscience, and Artificial Intelligence: A Volume in the Neuroscience of Neurodegenerative Disease Series*, stands as a pioneering initiative to confront this challenge head-on.

At its core, this book seeks to be more than just a compilation of research; it aims to be a transformative force, driving the field of translational neuroscience toward practical, tangible solutions. By focusing on the integration of neuroimaging, computational neuroscience, and artificial intelligence, this comprehensive work provides readers with a 360-degree view of neurodegenerative diseases. Through this multidimensional perspective, the book aims to illuminate unexplored connections, shed light on novel patterns, and inspire innovative strategies for diagnosis and treatment.

Through the collaboration of leading experts in neurology, neuroscience, computer science, and artificial intelligence (AI), this book offers an unmatched depth of knowledge, encompassing the latest research findings, breakthrough methodologies, and cutting-edge technologies. It serves as a beacon for researchers and clinicians alike, guiding them toward new frontiers in the battle against neurodegenerative disorders. From the practical implementation of AI in diagnostic settings to the utilization of computational neuroscience models to decipher the underlying mechanisms of neurodegeneration, the book leaves no stone unturned in its quest for progress.

The true strength of *Integrating Neuroimaging, Computational Neuroscience, and Artificial Intelligence* lies in its unique and transformative features. Firstly, it embraces an integrative approach that seamlessly weaves together insights from diverse fields, providing readers with a comprehensive understanding of neurodegenerative disorders in all their complexity. Secondly, the book places a strong emphasis on translational science, bridging the gap between research discoveries and tangible clinical applications. By highlighting the real-world implications of cutting-edge neuroscience, the book seeks to empower healthcare professionals to make a genuine impact on patient care.

Furthermore, this book looks beyond the present and gazes toward the future of neurodegenerative disease research and treatment. By envisioning personalized treatment approaches and innovative drug development strategies, the book stimulates critical discussions that can steer the field in groundbreaking directions. Through this visionary lens, *Integrating Neuroimaging, Computational Neuroscience, and Artificial Intelligence* is poised to become a catalyst for progress, inspiring collaborative efforts and accelerating the pace of discovery.

This ambitious book aspires to be more than a static collection of knowledge; it is envisioned as a dynamic force that will reshape the landscape of translational neuroscience and neurodegenerative disorder research. Through its interdisciplinary approach, emphasis on translational impact, and forward-looking vision, *Integrating Neuroimaging, Computational Neuroscience, and Artificial Intelligence* seeks to rally researchers, clinicians, and policymakers in a unified pursuit of better diagnostics, treatments, and ultimately, improved lives for those affected by neurodegenerative disorders.

Unique Features:

- Interdisciplinary integration: The book uniquely integrates diverse fields of research, from neuroimaging to computational neuroscience and AI, fostering a holistic approach that allows readers to grasp the complexities of neurodegenerative diseases from multiple angles.
- Translational focus: Unlike traditional neuroscience texts that may remain confined to theoretical concepts, this book emphasizes the translation of research findings into practical applications, ultimately benefitting patients and clinical practice.
- Real-world applications: By examining the implementation of AI and computational models in actual diagnostic settings, the book provides valuable insights into the practical implications of cutting-edge technologies, making it an essential resource for researchers, clinicians, and policymakers alike.
- Vision for the future: The book not only elucidates the current state of translational neuroscience but also outlines a vision for the future, exploring potential areas for further research and collaboration, including personalized treatments and drug development.
- Multidisciplinary collaboration: Through the collaboration of experts from diverse disciplines, the book encourages knowledge sharing and synergy, stimulating innovative approaches that may not arise within isolated silos of research.

This book is a call to action, inviting the scientific and medical communities to join forces in the pursuit of groundbreaking advancements in the fight against neurodegenerative diseases.

About the Editors

Dr. Indranath Chatterjee, Ph.D., is working as an Assistant Professor in the Department of Computing and Mathematics at Manchester Metropolitan University, UK. He is also working as an Adjunct Professor at the Woxsen University, India. He also holds the position of Director (International Collaborations) of The Korea Multimedia Society (KMMS), South Korea. Before that, he worked as an Assistant Professor at Tongmyong University, South Korea, for around five years and at JK Lakshmipat University, India, for around one year. He also worked as Research Director at Total Soft Bank, Pvt Ltd, South Korea, for one year. Dr. Chatterjee earned his Ph.D. in Computational Neuroscience from the University of Delhi, India. His research areas include computational neuroscience, schizophrenia, neuroimaging, and machine learning. He has authored and edited 14 books on computer science and neuroscience published by renowned international publishers. To date, he has published numerous research papers in international journals and conferences. He has also completed seven sponsored R&D projects as principal investigator (PI) from government and industries. He is a recipient of various global awards in neuroscience. Throughout his career, he has presented 26 keynote/plenary talks at various international conferences and seminars worldwide. He is currently serving as a chief section editor for a few renowned international journals and serving as a member of the advisory board and editorial board of various international journals and open-science organizations worldwide. He is an active professional member of the FENS, Belgium; ACM, USA; KMMS, Korea; OHBM, USA; ACNM, India; ALBA Network (Belgium); SONA, South Africa; and INCF, Sweden.

Dr. Nasrollah Moradikor, Ph.D., is currently Research Director and Head of Brain Aging and Dementia at the International Center for Neuroscience Research in Georgia. Dr. Nasrollah has demonstrated exceptional leadership skills in promoting applied neuroscience research and establishing fruitful collaborations among scientists worldwide. In 2021, Dr. Moradikor founded a new community platform named "Neuroscience Network" to support and develop neuroscience in the world. His ability to create a supportive and intellectually stimulating environment has been instrumental in promoting the development of aspiring neuroscientists. He has over 15 years of research and teaching experience, and he has chaired many national and international events in the area of neuroscience. Dr. Moradikor has been on the scientific advisory board and evaluation committees of several institutions abroad. He has published more than 100 scientific papers, and he currently serves as an editor, editorial board member, and section editor for several reputed journals in the field of neuroscience. His contributions to the scientific community are evident through his publications, as well as more than 15 years of editorial expertise. He is an active professional member of IBRO, FENS, EBBS, ISN, ESN, and MDS.

Contributors

Adeel Ahmed
Institute for Intelligent Research
Tbilisi, Georgia

Niaz Ahmed
National Tuberculosis Control Program
Islamabad, Pakistan

Gideon S. Alex
University of Port Harcourt
Rivers State, Nigeria

Abid Ali
Harbin Medical University
Harbin, PR China

Ashish
Department of Computer Science Engineering
Bhagwan Parshuram Institute of Technology
New Delhi, India

Madan Mohan Tito Ayyalasomayajula
Aspen University
Plano, CO, USA

Indranath Chatterjee
Department of Computing and Mathematics
Manchester Metropolitan University
Manchester, United Kingdom

Rajendra Prasad Chatterjee
Indian Council of Medical Research–National Institute of Cholera and Enteric Diseases
Kolkata, India

Shilpa Chatterjee
Department of Pharmaceutical Technology
Adamas University
Kolkata, India

Sejuti Ray Chaudhary
Adamas University
Kolkata, India

Swesme Enyioma-Alozie
Baze University
Abuja, Nigeria

Akansha Gautam
Department of Computer Science
University of Delhi
Delhi, India

Suruchi Gautam
Department of Computer Science
University of Delhi
Delhi, India

Sourav Ghosh
Adamas University
Kolkata, India

Mushahid Hussain
University of Sindh
Jamshoro, Pakistan

Adejoke Elizabeth Memudu
Anatomy Department
Edo State University Uzairue
Edo State, Nigeria

Nasrollah Moradikor
Institute for Intelligent Research
Tbilisi, Georgia

Farnaz Norouziyan
Islamic Azad University Urmia Campus
Urmia, Iran

Emmanuel Kenechukwu Nwanama
Enugu State University of Science and Technology
Enugu, Nigeria

Baliqis Adejoke Olukade
Department of Molecular Medicine
University of South Florida
Tampa, FL, USA

Augustine Oviosun
Department of Anatomy
Edo State University
Uzairue, Nigeria

Monosiz Rahaman
Adamas University
Kolkata, India

Dhvanit Rajdeep
Caucasus International University
Tbilisi, Georgia, USA

Muhammad Liaquat Raza
Department of Infection & Prevention Control
Ministry of the National Guard - Health Affairs
Riyadh, Saudi Arabia

Himani Sharma
Department of Computer Science Engineering
Bhagwan Parshuram Institute of Technology
New Delhi, India

Aditya Azad Singh
Department of Computer Science Engineering
Bhagwan Parshuram Institute of Technology
New Delhi, India

Aman Pratap Singh
Department of Computer Science Engineering
Bhagwan Parshuram Institute of Technology
New Delhi, India

Muhammad Umar
Shenzen Institute of Advanced Technology
Shenzen, PR China

Abbreviations

AD	Alzheimer's disease
ADNI	Alzheimer's Disease Neuroimaging Initiative
AI	artificial intelligence
ALS	amyotrophic lateral sclerosis
CSF	cerebro-spinal fluid
DA	dopaminergic
FTLD	frontotemporal lobar degeneration
GWAS	genome-wide association studies
HTT	huntingtin
MRI	magnetic resonance imaging
NDDS	neurodegenerative diseases
NFTS	neurofibrillary tangles
NGS	next-generation sequencing
PD	Parkinson's disease
PET	positron emission tomography
PM	precision medicine
PMCA	protein misfolding cyclic amplification
PMI	precision medicine initiative
PRSs	polygenic risk scores
ROSMAP	Religious Orders Study and Memory and Aging Project
RT-QuIC	real-time quaking-induced conversion
SNpc	substantia nigra pars compacta
SPECT	single-photon emission CT
TWAS	transcriptome-wide association studies

1 Introduction to Translational Neuroscience of Neurodegenerative Disorders

Adejoke Elizabeth Memudu, Augustine Oviosun, Emmanuel Kenechukwu Nwanama, and Swesme Enyioma-Alozie

1.1 INTRODUCTION TO TRANSLATIONAL NEUROSCIENCE

Translational neuroscience (TN) has traditionally been conceptualized to being a "bench-to-bedside" model, whereby new technologies and therapeutic strategies were designed based on direct findings from laboratory research or preclinical trials in animal models (Davies et al., 2020). Translational neuroscience (TN) is defined as "a systematic, theory-driven approach aimed to develop and leverage basic and clinical neuroscientific knowledge to aid the development and optimization of clinical and public health interventions" (Horn et al., 2020). TN entails a close collaboration between basic and clinical neuroscientists directed toward increasing neuroscience knowledge diversity across various fields while advancing clinical practice and the innovative development of therapeutic measures to combat neurodegenerative and mental health disorders. Without translational neuroscience, decades of basic research on mechanisms of neurodegeneration and therapeutic interventions that avert neurodegeneration would not yield successful clinical application for the treatments of neurological disorders (Davies et al., 2020). The use of animal models to translate neuropathological changes equivalent in humans has provided basic insights into the interplay between the nervous and immune system against protein aggregation in cases of neurocognitive and motor disorders leading to the development of therapeutics for neurological disorders (Mortada et al., 2021), such as Parkinson's disease (PD) and Alzheimer's disease (AD) studied in animal models (Lear et al., 2022). Visual-cognitive functions are studied in animal models such as monkeys (Beauchamp et al., 2020; Chen et al., 2020). Marmoset models are used to study modulation of prefrontal cortex regions in response to emotional stimuli in model of depression and anxiety (Alexander et al., 2019; Wallis et al., 2019). This has resulted in novel perspectives to individualized precision medicine for mood disorders (Roberts, 2020). It is important to reiterate that translational neuroscience is pivotal to understanding neuropathological mechanisms for precision medicine in neurological disorders, such as AD, PD, multiple sclerosis (MS), Huntington's disease (HD), and amyotrophic lateral sclerosis (ALS) (Mahalmani et al., 2022), thereby advancing medical science through bridging the gap between basic research and clinical applications that turn scientific discoveries into practical treatments, interventions, and therapies in the healthcare system.

DOI: 10.1201/9781032711102-1

1.2 KEY SIGNIFICANCE OF TRANSLATIONAL NEUROSCIENCE IN BRIDGING THE GAP BETWEEN BASIC RESEARCH AND CLINICAL APPLICATIONS

The knowledge of the mechanism of neurodegeneration and neuroregeneration in neurological disorders is obtained through scientific evidences such as neuroimaging, neuroendocrine or psychoneuroimmune activity, and neurobehavioral activities highly applied in preclinical trials in basic research and clinical trials to validate therapeutics (Horn et al., 2020) which has helped in precision medicine for positive outcomes in healthcare. There are a number of critically important advantages of translational neuroscience in the treatment of neurodegenerative disease therapy some are:

a. **Improving patient care outcomes:** Translational neuroscience research aims to directly benefit patients through the development of innovative strategies, diagnostics, therapies, and interventions that will potentially improve patient outcomes, enhance quality of life, and even save lives.
b. **Advancing and sustains biomedical research:** Advancement of basic research that unravels fundamental insights into biology, genetics, and disease mechanisms into the alleviation of disease to improve healthcare and longevity is significant. However, basic research discovery needs to be translated into clinical trials which advances medical innovations and therapeutic measures to combat neurological disorders.
c. **Targeted therapies:** Translational neuroscience research enables the development of brain-specific targeted therapies based on a deeper understanding of neurodegenerative disease pathways and mechanisms, resulting in a more effective and personalized treatments, reducing side effects and improving patient response (Mahalmani et al., 2022).
d. **Bridging the gap between bench and bedside:** The translation of basic research findings into clinical applications helps bridge the gap between laboratory discoveries and real-world patient care, resulting in multidisciplinary collaboration between researchers, clinicians, and industry, which is essential for translating knowledge into practice (Mahalmani et al., 2022).
e. **Promoting interdisciplinary collaboration:** Translational neuroscience research encourages collaboration between scientists from various disciplines through interdisciplinary or multidisciplinary research which fosters innovation and accelerates the development of new medical solutions for major neurological disorders (De Maria Marchiano et al., 2021).
f. **Enhancing healthcare delivery:** Beyond new therapies, translational neuroscience research can potentiate innovative healthcare delivery and patient care such as an improved diagnostic tools for earlier detection of neurodegenerative diseases, better management strategies, and more effective and efficient healthcare management practices (Davies et al., 2020).
g. **Attracting funding and investment:** Translational neuroscience research often attracts better funding and investment from governments, philanthropic organizations, and the private sector based on the pools of multidisciplinary and diversified research team to unravel the mechanism of neurodegeneration while investigating

innovative strategies to manage and improve patient outcome and also enable multidisciplinary research ecosystems for an effective and efficient collaboration (Davies et al., 2020).

1.3 RELEVANCE OF TRANSLATIONAL NEUROSCIENCE TO UNDERSTANDING AND TREATING NEURODEGENERATIVE DISORDERS

Translational neuroscience, which explores basic research and multidisciplinary discoveries to clinical settings, is important to promote sustainable development goals for human health (Manjarrez et al., 2023).

Neurodegenerative disorders (NDs) such as AD, PD, HD, MS, etc. (Naqvi et al., 2020) result from a progressive degeneration of the neurons in the brain and spinal cord (Silveira et al., 2021. This culminates into abnormal neural activity, cell death, cognitive impairments, and increase in mortality rate in the aging population (Hanif et al., 2020). NDs is a global menace of public health concern based on an epidemiology health report in 2020, which revealed that the global population of individuals living with NDs has exceeded 55 million people and this number is projected to double every 20 years, reaching 78 million in 2030 and 139 million in 2050 (Hanif et al., 2020). Hence, there is an urgent need to curtail this menace through swift development of novel innovative medical solutions targeted at effective treatment of neurological disorders (De Maria Marchiano et al., 2021). Translation of scientific knowledge from basic research evidences of the mechanism of neurodegeneration being triggered by oxidative stress and inflammatory processes (Silveira et al., 2021) will help develop precision novel therapeutic interventions for NDs which are targeted at neuro-inflammatory processes associated with the activation of microglial, astrocytes, and secretion of pro-inflammatory mediators. Till date, most therapeutic measures for NDs are targeted at neuromodulatory effects; for example, AD patients are given drugs such as donepezil (acetyl-cholinesterase inhibitors) or galantamine (a nicotinic acetylcholine receptor modulators) or memantine (an N-Methyl-D-Aspartate [NMDA] receptor antagonists) to reduce disease progression, making therapeutic approaches palliative (Hanif et al., 2020) due to major hindrances in brain targeted drug delivery and discovery of drugs targeting neuropathologic mechanistic pathways. This has limited the available therapeutic agents for NDs, hence the urgent need to develop novel therapeutic strategies for the treatments of NDs (Davies et al., 2020; Hanif et al., 2020; Naqvi et al., 2020). Translational neuroscience is key to resolve therapeutic strategies for NDs by the application of knowledge acquired from basic neuroscience research into therapeutic interventions for treating human diseases such NDs (Davies et al., 2020). In translational neuroscience, "wet laboratory research" helps to identify and acknowledge the mechanism of NDs for targeted therapeutic interventions (Davies et al., 2020; Hanif et al., 2020) done using animal brain models for preclinical trials in laboratory settings now evolving to human stem cells and postmortem brain samples (Davies et al., 2020). Commonly applied experimental animal models for preclinical studies include rodents such as mice and rats though they lack the intricacies of the human brain and hence do not demonstrate the complexities of human pathology evidently ascribed to their distant similarity with neuropathology in human postmortem brain samples (Götz et al., 2018; Davies et al., 2020). However, these animal models are continually undergoing improvement to model-specific neurological disease processes using models such as transgenic mice and rats (Götz et al., 2018) obtained through injection of pathogenic risk factors of NDs, for example, studies involving antisense oligonucleotide remedial treatment for spinal muscular atrophy (Corey, 2017). Another innovative model in translational neuroscience is the use of human-induced pluripotent stem cell technology to study the complexity of NDs disease progression and response to therapeutic interventions; human-induced pluripotent stem cell-derived brain cells and organoids model provide a promising measure for evaluating the intricacies of inherited disorders although it has been documented to include the intricacies of the human brain, similar to rodents as vertebrate models for studying specific neuropathological conditions (Davies et al., 2020). Zebrafish has also proven to be a model in consolidating the genotypes and phenotypes of a broad spectrum of human neurological conditions; hence, it is a preferred laboratory model for assessing neuropathophysiological mechanisms underscoring vast neurological diseases (Rajanikant and Kalthur, 2022). According to Manjarrez et al. (2023), translational neuroscience helps to bridge the gap between "bench and bed side" with a significant impact on the understanding of brain function and potential clinical uses application in therapeutic discoveries. Translational neuroscience is vital in understanding and treating NDs, such as AD, PD, HD, and ALS (Davies et al., 2020; Hanif et al., 2020) resulting in new drug discoveries, identification of brain targeted therapeutic mechanisms, and biomarkers discoveries (Dawson et al., 2018). Translational neuroscience supports the development of personalized medicine tailored treatments to an individual's specific disease profile which makes the treatment more effective with less side effects or harmful interventions (Dawson et al., 2018; Strafella et al., 2018). From the bench to the bedside, precision medicine can be made available through the development of collaborative partnerships between academic institutions, hospital centers, and highly skilled professionals in addition to involving policy makers in industries (Strafella et al., 2018; Davies et al., 2020). Furthermore, through translational neuroscience, clinicians can collaborate with basic scientists to identify novel biomarkers for the earlier detection and monitoring of the progression of NDs. These biomarkers can be specific proteins, genetic markers, or neuroimaging findings which enable earlier intervention and more targeted treatments (Tanaka et al., 2022, 2023). These biomarkers of ND detect the process of neurodegeneration prior to clinical presentation needed to improve clinical trial design and provide relevant outcomes. Translational neuroscience helps unravel the correlation between aging, gender, pathophysiology, and clinical symptoms of some neurological disorders, using preclinical models leading to the discovery of histone deacetylases (HDACs) known as sirtuins (SIRT1–SIRT7), a special enzyme dependent on NAD+ levels which mediates neurodegeneration in AD and aging (Tanaka et al., 2022). Research using surface plasmon resonance (SPR) technology has helped to demonstrated a correlation between serum levels of sirtuins (SIRT1–SIRT7) proteins and stage of neurodegenerative diseases (NDs). Specifically, a decline in SIRT1 and SIRT3 mRNA/protein levels in the serum of Alzheimer's disease (AD) patients has been identified as biomarker for mild cognitive impairment in AD. The advancements in neuroimaging modalities such as magnetic resonance imaging (MRI), positron emission tomography (PET) scans, and cerebrospinal fluid studies, in addition to computational models, offer important diagnostic and surveillance data for identifying and monitoring the progression of neurodegenerative diseases (Specht, 2020). The current frontier in the treatment strategy for neurological patients involves nano agents as nanoprobes for clinical markers applicable to multimodal neuroimaging (Bonanno and Calabro, 2023). One of the most popular neuroimaging techniques uses lateralization and localization of linguistic domains, which is crucial knowledge for treating epileptic patients (Specht, 2020). Concerning new drug discovery in the management of depression, 828 antidepressants were developed at the end of 2019, out of which only 292 are reported to be active and effective, while more than half were annulled. This indicates the need to keep developing new drugs in the field of depression (Li et al., 2021). Based on various preclinical studies done for new drug discovery to manage depression, two new antidepressants that act as N-methyl-D-aspartate (NMDA) receptor antagonist were discovered to help avert symptoms of depressions, which results in conducting a large number of prelusive clinical studies (Zheng et al., 2020). Several randomized controlled trials studies show that derivate of the ketamine such as esketamine is productive in depression or remedy-resistant depression (Popova et al., 2019). The collaborative role of basic and clinical neuroscience to advance science policy in the industry such as the Food and Drug Administration (FDA) helped to ratify two novel antidepressants: esketamine and esmolamine for refractory depression and bresanolone for postpartum depression (Cristea and Naudet, 2019). According to Charles et al. (2012), deep brain stimulation (DBS) of subthalamic nucleus used in animal models resulted in an enhanced viability of dopaminergic neurons in advanced PD. There are increasing scientific reports to corroborate a DBS neuroprotective role to modify the pathophysiology of AD, PD, and epilepsy in preclinical models (Fischer et al., 2017). A nonrandomized trial, carried out by Ngoga et al. (2014) involving bilateral subthalamic nucleus deep brain stimulation (STN-DBS) in patients with severe PD refractor, revealed an improved motor function, motor fluctuations, health-related quality of life, and a reduced medication usage and drug-induced dyskinesia in PD patients compared to controls exposed to the medication alone. The trends in global incidence of NDs are predicated to increase globally by 2040; hence, there is a need for drastic measures through translational neuroscience best practices for better public health policy in healthcare facilities directed toward personalized medicines (Li et al., 2023).

1.4 OVERVIEW OF NEURODEGENERATIVE DISORDERS

1.4.1 Definition and Characteristics of Neurodegenerative Disorders

NDs such as AD, PD, HD, MS, prion disease, ALS, motor neuron disease, spinal muscular atrophy, and spinocerebellar ataxia are collectively termed NDs (Kovacs, 2019; Naqvi et al., 2020). They occur due to progressive degeneration of the neurons in the brain and spinal cord leading to abnormalities in neural activity that culminates in cell death and cognitive impairments (Hanif et al., 2020). Neuron loss and reactive astrogliosis are the hallmark for neurodegeneration neuropathology (Kovacs, 2019; Memudu et al., 2024). Certain (sub)populations of nervous system cells that are very important in movement, coordination, strength, sensation, and cognition are present at the late onset of progressive damage (Lamptey et al., 2022; Pathak et al., 2022; Pozojevic and Spielmann, 2023).

1.4.2 Prevalence and Societal Impact of Neurodegenerative Disorders

Neurological disorders and NDs are a major health problem worldwide, which accounts for an increase in global mortality rate (Ding et al., 2022). The increased prevalence of NDs is associated with population growth, aging, genetics, and environmental predisposing factors making ND a global burden worldwide (Huang et al., 2023). In 2019 alone, neurological disorders were responsible for nearly 10 million deaths and 349 disability-adjusted life years (DALYs) lost globally (Ding et al., 2022; Zhang et al., 2023). In 2021, approximately 3.4 billion people were living with a neurological disorder, which accounts for 11.1 million mortality and 443 million DALYs, hence making NDs the major contributor to global DALYs (Huang et al., 2023; Steinmetz et al., 2024). In East Africa, NDs are highly prevalent, impacting the rate of morbidity, quality of life, and life expectancy (Onohuean et al., 2022). Despite ongoing rapid research interventions, most patients with neurodegenerative diseases (NDs) are diagnosed at a late stage, when significant neuronal damage has already occurred. This delays optimal healthcare interventions, highlighting the urgent need for early detection strategies before the onset of neuropathological symptoms (Huang et al., 2023). NDs impact life expectancy significantly resulting in extrapyramidal and pyramidal movement disorders and cognitive or behavioral disorders which alter quality of life of the individual living with NDs (Pathak et al., 2022).

1.4.3 Mechanism of Neurodegeneration

Neurodegeneration is caused by changes in the brain through oxidative stress-linked neuroinflammation, mitochondrial dysfunction, and abnormal protein homeostasis (Andreone et al., 2019). Age, genetic variations, traumatic injury, oxidative stress, inflammation, toxic exposure, and stress are a few of the risk variables that exist generally (Hou et al., 2019). Although a well-known genetic mutation may account for the origin of NDs in certain cases, idiopathic or even environmental variables may occur in some cases (Pang et al., 2019). Several studies have linked oxidative stress to the neuropathology of neurodegenerative diseases (NDs) due to an abnormal increase in the release of reactive oxygen species (ROS). This excessive ROS triggers mitochondrial dysfunction, leading to excitotoxic neuronal cell death in NDs (Ohnose et al., 2022). Protein accumulations and anatomical alterations are commonly used to characterize neurodegenerative diseases as a result of processes linked to the progressive dysfunction and death of neurons, such as oxidative stress, neuroinflammation, proteotoxic stress, abnormalities in the ubiquitin-proteasomal and autophagosomes or lysosomal systems, and programmed cell death (Lamptey et al., 2022; Pathak et al., 2022). Importantly, it is necessary to note protein abnormalities that connote NDs could occur before the onset of clinical symptoms or manifestations (Kovacs, 2019). PD is a rapid motor disorder (Yang et al., 2020; Bloem et al., 2021; Huynh et al., 2021). Epidemiology shows that 5% of people have genetic etiology for PD, characterized by Lewy bodies and Lewy neurites, which are neuronal accretions containing the protein α-synuclein, neuropathological expression of PD (Shahmoradian et al., 2019). AD is the common form of dementia and characterized by the loss of cognitive function attributed to genetic risk factors such as dominant variations in amyloid precursor protein (APP) (which codes for amyloid precursor protein), PSEN1, PSEN2 (which code for presenilin 1 and 2, respectively), APOE, TREM2, and MS4 resulting in the formation of neurofibrillary tangles and plaques containing β-amyloid (Knopman et al., 2021). According to recent research on the pathogenesis of AD, tau and amyloid-beta (Aβ) protein accumulation is required in AD neuropathological progression (Pickett et al., 2019). However, cellular homeostatic mechanism through the antioxidant system helps to repair this neuropathological progression via endogenously generated antioxidant compound that identifies generated ROS and mobs its off the cell milieu (Misrani et al., 2021). Anti-inflammatory agents help reverse the pathophysiology of neurodegenerative diseases (NDs) by activating microglia, which clear amyloid-beta plaques through phagocytosis, serving as a mechanism of neuroprotection and promoting an anti-inflammatory response (Yousefi et al., 2021; Onohuean et al., 2022). Translational neuroscience research is aimed to understand cellular, molecular, and pharmacological mechanisms to avert the neuropathology of NDs with the ultimate aim to implement in clinical trials (Wareham et al., 2022). A number of research have been conducted to understand the etiology of various NDs for effective precision medicine in clinical trials. For instance Smajić et al. (2022) carried out a single-nucleus sequencing (snRNA-seq) of frozen human postmortem midbrains (six PD patients and five controls). This study discovered that idiopathic PD midbrain has an increased expression of CADPS2, which codes for a calcium-dependent activator, low amount of TH; a marker gene that encodes tyrosine hydroxylase needed for the translation of tyrosine to dopamine, increased astrocytes, and microglia which demonstrate an increase in the genes linked to the inflammatory pathway (*IL1B, GPNMB, HSP90AA1*) due to oxidative stress. This aforementioned study involves cultures of primary cell types to study idiopathic PD progression mechanism (Smajić et al., 2022). Furthermore, the application of innovative technology such as genome-wide association study (GWAS) will help translational neuroscience for guided precision medicine (Agarwal et al., 2020; Bryois et al., 2020). In postmortem AD brains, Mathys et al. (2019) examined prefrontal cortex samples from 48 patients and discovered that multiple cell types had perturbations in myelination-related processes with increased microglia which is taken as a risk factor for APOE. Increased expression of microglia correlates with the discoveries by Grubman et al. (2019) and Lau et al. (2020). This indicates that microglia expression in AD pathology is similar to endothelial cell activation in neurodegeneration conserved in both humans and animals (Lau et al., 2020) which results in neuronal malfunction and death in the cerebral cortex (Lim et al., 2022). In Huntington disease, using HD-affected animal model (Lim et al., 2022), it was observed that mutant huntingtin (HTT) released mitochondrial RNA, the activated innate immunological signaling, while downregulating the mitochondrial oxidative phosphorylation pathway (Lee et al., 2020). Astrocytes proliferation reveals transcriptional phenotypes associated with HD which exhibits elevated heat shock and metallothionein genes, which enhance glial fibrillary acidic protein, a reactive astrocyte marker (Al-Dalahmah et al., 2020). This correlates with research that demonstrated the expression of the gene in microglia and astrocytes that regulate angiogenesis and blood vessel endothelial cell migration along the blood–brain barrier in NDs (Garcia et al., 2022; Malaiya et al., 2021).

1.5 IMPORTANCE OF TRANSLATIONAL NEUROSCIENCE OF NEURODEGENERATIVE DISORDERS

1.5.1 Increasing Burden of Neurodegenerative Disorders in Aging Populations Amid the Absence of Effective Treatments' Strategies

The basic challenge that is paramount in translational neuroscience is seen in biological complexity of NDs and cell-targeted drug delivery attributed to the peculiar structure of the blood–brain barrier. NDs are difficult to diagnose, and drugs do not easily get into the brain. Furthermore, these disorders are mostly diagnosed in an aging patient population that has been exposed to several other conditions that may be altogether life threatening (Zeiss, 2017). Regrettably, the origination and development of most neurodegenerative diseases such as AD and PD are not yet understood, and there is paucity of diagnostic markers for early detection as well as lack of efficacious remedies for these diseases. Most cases of

AD are diagnosed usually after the symptoms have become visible, and it occurs more than 10 years after early pathological alterations on the brain tissue; therefore, reversal or treatment becomes almost impossible (Zheng and Chen, 2022).

1.5.2 Urgent Need for Innovative Approaches to Diagnosis, Treatment, and Prevention through Discovery of New Biomarkers

Translational research in neuroscience is increasingly focusing on the analysis of multimodal data in order to account for the biological complexity of suspected disease mechanisms (Sun et al., 2018; Li et al., 2021). An epidemiology report on ND global prevalence predicted that by 2050, the ratio will possibly double to one in six, from the reported 1 out of every 11 world populations being more than 65 years in 2019 (Zheng and Chen, 2022). The prevalence of NDs increases with the growth of the global aging population. An alarming projection of a rise in dementia cases in the developed world to 36.7 million by the year 2050 as against the previous projection from 13.5 million in 2000 to 21.2 million in 2025 (Zheng and Chen, 2022) is another reason for an urgent call for collaborative efforts toward developing innovative approaches for early diagnosis, effective treatments, and creative mechanisms for the prevention of neurodegenerative diseases. The identification of early biomarkers expressed in ND progression for earlier detection and diagnosis of NDs is systematic evidence, which translates research to clinical trials and clinical practice (Kapur et al., 2012). The process of managing NDs is conceptualized with biomarker discovery and development, through preliminary proof-of-concept studies of the potential biomarker (FDA, 2018). New biomarker discovery occurs through the collection of clinical samples, such as blood, and the analysis of plasma and serum (the primary source for biomarker discovery) by mass spectrometry to identify key proteins in cellular interaction regulating many neurophysiological processes and neuropathological conditions forms a good predictor of the cellular condition of brain tissue (Haqqani and Stanimirovic, 2013) in neurological diseases. One method for early diagnosis of PD is the isolation and identification of cells secretome in the plasma sample of patients with PD (Anjo et al., 2020), while early diagnostic intervention for AD involves the use of optical coherence tomography (OCT), a noninvasive technology to examine retina microcirculation (Zheng and Chen, 2022). The use of biological fluid and anatomic structures will aid as rapid innovative targets to identify NDs and provide better understanding of the molecular, cellular, and anatomical mechanisms of NDs through an interdisciplinary diverse team of both clinicians and basic researchers interested in "bedside-to-bench" approach for effective health outcomes in disease diagnosis and treatment.

1.5.3 Current Approaches and Next Step in Translational Neuroscience Research

Translational neuroscience intersects the application of basic and clinical neuroscience to understand the mechanisms of neuronal activity in health and diseases directed toward creating an innovative model for early diagnosis and discovery of positive therapeutic outcomes in precision medicine for people living with ND (Woo et al., 2017). The current approach involves innovative technologies for the early detection of NDs using functional neuroimaging which has generated great hope in revolutionizing basic and clinical knowledge of NDs (Mather et al., 2013), with few models from functional neuroimaging integrated into clinical diagnosis (Kapur et al., 2011). Translational neuroimaging has provided valuable insights by identifying brain features through the measurement of regional neural activity, helping predict health-related outcomes for NDs such as major depressive disorder and cognitive disorders (Davies et al., 2020). Early use of traditional brain mapping methods in translational neuroimaging has been key in identifying and understanding functional processes within targeted brain regions of interest (Stoyanov, 2017). This has advanced to the use of standard parametric mapping scales to create whole-brain maps to mark specific brain regions vital in expression clinical manifestations of ND disorders for example researchers unravel the correlation between subgenual ventral cingulate cortex and mood disorder like depression (Ibrahim et al., 2022), thalamus and periaqueductal gray with chronic pain (Luo et al., 2021), basal ganglia with obsessive-compulsive disorder (Parmar and Sarkar, 2016), and subthalamic nucleus with PD (Santos et al., 2023). These associations are helpful in clinical trials targeting specific brain regions for neuromodulation therapies, such as deep-brain stimulation and neurofeedback, and have been proved to be effective in some neurological disorders (Bichsel et al., 2021; Camacho-Conde et al., 2022; Rissardo et al., 2023; Ubeda Matzilevich et al., 2024).

1.5.4 Challenges in Translational Neuroscience Research

The increasing demographical projections of dementia from 13.5 million in 2000 to 21.2 million in 2025 and the predicted increase of 36.7 million in 2050 in the midst of limited resource or disease models to discover new therapeutic strategies for precision medication (Zheng and Chen, 2022) is a challenge. In order to completely illuminate the mechanisms of neurodegenerative diseases, there is a need to make available satisfactory model systems such as authentic animal models that play essential roles in the comprehension of pathways and interactions of diseases and body systems since there is always a difference between biochemical and behavioral phenotypes of diseases in animal models and humans based on the complexity of ND disease progression in humans (Davies et al., 2020). Using neuroimaging modalities such as MRI BOLD signals of brain functions has been limited with some challenges such some limitations in correlating neuroimaging results between animal models and human clinical trials for neurodegeneration mechanisms and therapeutics (Ibáñez et al., 2018). Genetic scientists and clinical researchers need to leverage on molecular biology approaches, such as polymerase chain reaction (PCR), and next-generation sequencing (NGS) to identify genes or changes in an individual's genetic makeup that underlie common neurodegenerative conditions such as AD, PD, and ALS using innovative technological advances that help to accelerate the discovery and understanding of early diagnostic markers that may predict disease prior to symptom onset, which is crucial for more personalized treatment strategies (Zheng and Chen, 2022). Biobanking of biological specimen is a great challenge to promote the knowledge of various neurological diseases, applicable to translational research. This is linked with the unavailability or lack of easy access to brain samples deposited in biobanks, as facility for collecting biospecimens and associated data, required to translate experimental findings into clinical trials (Giannella et al., 2023). It is important to discuss the role of biobanking for translational research on neurodegenerative diseases through standardization in the preanalytical phase, ethics, and legal consideration for patient data protection (Giannella et al., 2023). Science advocacy and education are integral in promoting clinical trial advancement through the application of modern innovative technologies such as neuroimaging techniques, stem cell biotechnology, exosome technology, and nanomedicine technology that will translate into rapid innovation in medicinal therapy development through large peer or cohort clinical trials that aim to bridge basic science with clinical applications in healthcare facilities (Leonelli, 2019; Zheng and Chen, 2022).

1.6 CONCLUDING REMARKS

The translational neuroscience of NDs is a dynamic, rapidly evolving field of scientific research that will revolutionize NDs precision medicine with immense potential for improving patient outcomes. Though faced with some limitations in the midst of its promising results, these collaborative efforts among researchers, clinicians, and industry partners to integrate preclinical research with clinical trials and patient care through innovative technologies, such as advanced imaging techniques, genetics, biomarker discovery, and drug development, would significantly accelerate the translation of emerging scientific discoveries into clinical practice resulting in an effective personalized medicine that will improve the quality of life for people living with NDs.

REFERENCES

Agarwal, D., Sandor, C., Volpato, V., Caffrey, T.M., Monzón-Sandoval, J., Bowden, R., Alegre-Abarrategui, J., Wade-Martins, R. and Webber, C. (2020). A single-cell atlas of the human substantia nigra reveals cell-specific pathways associated with neurological disorders. *Nature Communication*, *11*, 4183.

Al-Dalahmah, O., Sosunov, A.A., Shaik, A., Ofori, K., Liu, Y, Vonsattel, J.P., Adorjan, I., Menon, V., & Goldman, J.E. (2020). Single-nucleus RNA-seq identifies Huntington disease astrocyte states. *Acta Neuropathologica Communication*, *8*, 19.

Alexander, L., Gaskin, P.L., Sawiak, S.J., Fryer, T.D., Hong, Y.T., Cockcroft, G., Clarke, H.F., & Roberts, A.C. (2019). Fractionating blunted reward processing characteristic of anhedonia by over-activating primate subgenual anterior cingulate cortex. *Neuron*, *101*(2), 307–320.

Andreone, B.J., Larhammar, M., & Lewcock, J.W. (2019). Cell death and neurodegeneration. *Cold Spring Harbor, Perspectives in Biology*, *12*, a036434. https://doi.org/10.1101/cshperspect.a036434

Anjo, S.I., dos Santos, P.V., Rosado, L. Baltazar, G., Baldeiras, I., Pires, D., Gomes, A., Januário, C., Castelo-Branco, M., Grãos, M., & Manadas, B. (2020). A different vision of translational research in biomarker discovery: A pilot study on circulatory mitochondrial proteins as Parkinson's disease potential biomarkers. *Translational Neurodegeneration*, *9*, 11. https://doi.org/10.1186/s40035-020-00188-0

Beauchamp, M.S., Oswalt, D., Sun, P., Foster, B.L., Magnotti, J.F., Niketeghad, S., Pouratian, N., Bosking, W.H., & Yoshor, D. (2020). Dynamic stimulation of visual cortex produces form vision in sighted and blind humans. *Cell*, *181*(4), 774–783.

Bichsel, O., Stieglitz, L.H., Oertel, M.F., Baumann, C.R., Gassert, R., & Imbach, L.L. (2021). Deep brain electrical neurofeedback allows Parkinson patients to control pathological oscillations and quicken movements. *Scientific Reports*, *11*(1), 7973. https://doi.org/10.1038/s41598-021-87031-2

Bloem, B.R., Okun, M.S., & Klein, C. (2021). Parkinson's disease. *Lancet*, *397*(10291), 2284–2303.

Bonanno, M., & Calabrò, R.S. (2023). Bridging the gap between basic research and clinical practice: The growing role of translational neurorehabilitation. *Medicines*, *10*(8), 45.

Bryois, J., Skene, N.G., Hansen, T.F., Kogelman, L.J.A., Watson, H.J., Liu, Z., Eating Disorders Working Group of the Psychiatric Genomics Consortium, International Headache Genetics Consortium, 23andMe Research Team, Brueggeman, L., Breen, G., Bulik, C.M., Arenas, E., Hjerling-Leffler, J., & Sullivan, P.F. (2020). Genetic identification of cell types underlying brain complex traits yields insights into the etiology of Parkinson's disease. *Nature Genetics*, *52*, 482–493.

Camacho-Conde, J.A., Gonzalez-Bermudez, M.D.R., Carretero-Rey, M., & Khan, Z.U. (2022). Brain stimulation: A therapeutic approach for the treatment of neurological disorders. *CNS Neuroscience & Therapeutics*, *28*(1), 5–18. https://doi.org/10.1111/cns.13769

Charles, P.D., Dolhun, R.M., Gill, C.E., Davis, T.L., Bliton, M.J., Tramontana, M.G., Salomon, R.M., Wang, L., Hedera, P., Phibbs, F.T., Neimat, J.S., & Konrad, P.E. (2012). Deep brain stimulation in early Parkinson's disease: Enrollment experience from a pilot trial. *Parkinsonism & Related Disorders*, *18*(3), 268–273.

Chen, X., Wang, F., Fernandez, E., & Roelfsema, P.R. (2020). Shape perception via a high-channel-count neuroprosthesis in monkey visual cortex. *Science*, *370*(6521), 1191–1196.

Corey, D.R. (2017). Nusinersen, an antisense oligonucleotide drug for spinal muscular atrophy. *Nature Neuroscience*, *20*(4), 497–499. https://doi.org/10.1038/nn.4508

Cristea, I.A., & Naudet, F. (2019). US Food and Drug Administration approval of esketamine and brexanolone. *The Lancet Psychiatry*, *6*(12), 975–977. https://doi.org/10.1016/S2215-0366(19)30292-5

Davies, C., Hamilton, O.K.L., Hooley, M., Ritakari, T.E., Stevenson, A.J., & Wheater, E.N.W. (2020). Translational neuroscience: The state of the nation (a PhD student perspective). *Brain Communications*, *2*(1), fcaa038. https://doi.org/10.1093/braincomms/fcaa038

Dawson, T.M., Golde, T.E., & Lagier-Tourenne, C. (2018). Animal models of neurodegenerative diseases. *Nature Neuroscience*, *21*(10), 1370–1379.

De Maria Marchiano, R., Di Sante, G., Piro, G., Carbone, C., Tortora, G., Boldrini, L., Pietragalla, A., Daniele, G., Tredicine, M., Cesario, A., Valentini, V., Gallo, D., Babini, G., D'Oria, M., & Scambia, G. (2021). Translational research in the era of precision medicine: Where we are and where we will go. *Journal of Personalized Medicine*, *11*(3), 216. https://doi.org/10.3390/jpm11030216

Ding, C., Wu, Y., Chen, X., Chen, Y., Wu, Z., Lin, Z., Kang, D., Fang, W., & Chen, F. (2022). Global, regional, and national burden and attributable risk factors of neurological disorders: The global burden of disease study 1990-2019. *Frontiers in Public Health*, *10*, 952161. https://doi.org/10.3389/fpubh.2022.952161

FDA. (2018). *Biomarker qualification: evidentiary framework*. FDA Center for Drug Evaluation and Research.

Fischer, D.L., Kemp, C.J., Cole-Strauss, A., Polinski, N.K., Paumier, K.L., Lipton, J.W., Steece-Collier, K., Collier, T.J., Buhlinger, D.J., & Sortwell, C.E. (2017). Subthalamic nucleus deep brain stimulation employs trkB signaling for neuroprotection and functional restoration. *Journal of Neuroscience and Neurological Disorders*, *37*, 6786–6796.

Garcia, F.J., Sun, N., Lee, H., Godlewski, B., Mathys, H., Galani, K., Zhou, B., Jiang, X., Ng, A.P., Mantero, J., Tsai, L.H., Bennett, D.A., Sahin, M., Kellis, M., & Heiman, M. (2022). Single-cell dissection of the human brain vasculature. *Nature*, *603*, 893–899.

Giannella, E., Notarangelo, V., Motta, C., & Sancesario, G. (2023). Biobanking for neurodegenerative diseases: Challenge for translational research and data privacy. *The Neuroscientist*, *29*(2), 190–201. https://doi.org/10.1177/10738584211036693

Götz, J., Bodea, L.G., & Goedert, M. (2018). Rodent models for Alzheimer disease. *Nature Reviews Neuroscience*, *19*(10), 583–598. https://doi.org/10.1038/s41583-018-0054-8

Grubman, A., Chew, G., Ouyang, J.F., Sun, G., Choo, X.Y., McLean, C. Simmons, R. K., Buckberry, S., Vargas-Landin, D.B., Poppe, D., Pflueger, J., Lister, R., Rackham, O.J.L., Petretto, E., & Polo, J.M. (2019). A single-cell atlas of entorhinal cortex from individuals with Alzheimer's disease reveals cell-type-specific gene expression regulation. *Nature Neuroscience*, *22*, 2087–2097.

Hanif, S., Muhammad, P., Chesworth, R., Rehman, F.U., Qian, R.J., Zheng, M., & Shi, B.Y. (2020). Nanomedicine-based immunotherapy for central nervous system disorders. *Acta Pharmacologica Sinica*, *41*, 936–953. https://doi.org/10.1038/s41401-020-0429-z

Haqqani, A.S., & Stanimirovic, D.B. (2013). Prioritization of therapeutic targets of inflammation using proteomics, bioinformatics, and in silico cell-cell interactomics. *Methods in Molecular Biology*, *1061*, 345–360.

Horn, S.R., Fisher, P.A., Pfeifer, J.H., Allen, N.B., & Berkman, E.T. (2020). Levers and barriers to success in the use of translational neuroscience for the prevention and treatment of mental health and promotion of well-being across the lifespan. *Journal of Abnormal Psychology*, *129*(1), 38–48. https://doi.org/10.1037/abn0000465

Hou, Y., Dan, X., Babbar, M., Wei, Y., Hasselbalch, S.G., Croteau, D.L., & Bohr, V.A. (2019). Ageing as a risk factor for neurodegenerative disease. *Nature Reviews Neurology*, *15*, 565–581. https://doi.org/10.1038/s41582-019-0244-7

Huang, Y., Li, Y., Pan, H., & Han, L. (2023). Global, regional, and national burden of neurological disorders in 204 countries and territories worldwide. *Journal of Global Health*, *13*, 04160. https://doi.org/10.7189/jogh.13.04160

Huynh, B., Fu, Y., Kirik, D., Shine, J.M., & Halliday, G.M. (2021). Comparison of locus coeruleus pathology with nigral and forebrain pathology in Parkinson's disease. *Movement Disorders*, *36*, 2085–2093.

Ibáñez, A., Sedeño, L., García, A.M., Deacon, R.M.J., & Cogram, P. (2018). Editorial: Human and animal models for translational research on neurodegeneration: Challenges and opportunities from South America. *Frontiers in Aging Neuroscience*, *10*, 95. https://doi.org/10.3389/fnagi.2018.00095

Ibrahim, H.M., Kulikova, A., Ly, H., Rush, A.J., & Sherwood Brown, E. (2022). Anterior cingulate cortex in individuals with depressive symptoms: A structural MRI study. *Psychiatry Research: Neuroimaging*, *319*, 111420. https://doi.org/10.1016/j.pscychresns.2021.111420

Kapur, S., Phillips, A.G., & Insel, T.R. (2012). Why has it taken so long for biological psychiatry to develop clinical tests and what to do about it? *Molecular Psychiatry*, *17*(12), 1174–1179. https://doi.org/10.1038/mp.2012.105

Knopman, D.S., Amieva, H., Petersen, R.C., Chételat, G., Holtzman, D.M., Hyman, B.T., Nixon, R.A. and Jones, D.T. (2021). Alzheimer disease. *Nature Reviews Disease Primers*, *7*(1), 33.

Kovacs, G.G. (2019). Molecular pathology of neurodegenerative diseases: Principles and practice. *Journal Clinical Pathology*, *72*, 725–735.

Lamptey, R.N.L., Chaulagain, B., Trivedi, R., Gothwal, A., Layek, B., & Singh, J. (2022). A review of the common neurodegenerative disorders: Current therapeutic approaches and the potential role of nanotherapeutics. *International Journal Molecular Science*, *23*, 1851. https://doi.org/10.3390/ijms23031851

Lau, S.-F., Cao, H., Fu, A.K.Y., & Ip, N.Y. (2020). Single-nucleus transcriptome analysis reveals dysregulation of angiogenic endothelial cells and neuroprotective glia in Alzheimer's disease. *Proceedings of the National Academy of Sciences*, *117*, 25800–25809.

Lear, A., Baker, S.N., Clarke, H.F., Roberts, A.C., Schmid, M.C., & Jarrett, W. (2022). Understanding them to understand ourselves: The importance of NHP research for translational neuroscience. *Current Research in Neurobiology*, *3*, 100049.

Lee, H., Fenster, R.J., Pineda, S.S., Gibbs, W.S., Mohammadi, S., Davila-Velderrain, J., Garcia, F.J., Therrien, M., Novis, H.S., Gao, F., Wilkinson, H., Vogt, T., Kellis, M., LaVoie, M.J., & Heiman, M. (2020). Cell type-specific transcriptomics reveals that mutant huntingtin leads to mitochondrial RNA release and neuronal innate immune activation. *Neuron*, *107*, 891–908.e8.

Leonelli, S. (2019). Data governance is key to interpretation: Reconceptualizing data in data science. *Harvard Data Science Review*, *1*(1). https://doi.org/10.1162/99608f92.17405bb6

Li, D.P., Han, Y.X., He, Y.S., Wen, Y., Liu, Y.C., Fu, Z.Y., Pan, H.F., & Cao, F. (2023). A global assessment of incidence trends of autoimmune diseases from 1990 to 2019 and predicted changes to 2040. *Autoimmunity Reviews*, *22*(10), 103407. https://doi.org/10.1016/j.autrev.2023.103407

Li, Z., Ruan, M., Chen, J., & Fang, Y. (2021). Major depressive disorder: Advances in neuroscience research and translational applications. *Neuroscience Bulletin*, *37*, 863–880. https://doi.org/10.1007/s12264-021-00638-3

Lim, R.G., Al-Dalahmah, O., Wu, J., Gold, M.P., Reidling, J.C., Tang, G., Adam, M., Dansu, D.K., Park, H.J., Casaccia, P. and Miramontes, R. (2022). Huntington disease oligodendrocyte maturation deficits revealed by single-nucleus RNAseq are rescued by thiamine-biotin supplementation. *Nature Communication*, *13*, 7791.

Luo, H., Huang, Y., Green, A.L., Aziz, T.Z., Xiao, X., & Wang, S. (2021). Neurophysiological characteristics in the periventricular/periaqueductal gray correlate with pain perception, sensation, and affect in neuropathic pain patients. *NeuroImage: Clinical*, *32*, 102876. https://doi.org/10.1016/j.nicl.2021.102876

Mahalmani, V., Sinha, S., Prakash, A., & Medhi, B. (2022). Translational research: Bridging the gap between preclinical and clinical research. *Indian Journal of Pharmacology*, *54*(6), 393–396. https://doi.org/10.4103/ijp.ijp_860_22

Malaiya, S., Cortes-Gutierrez, M., Herb, B.R., Coffey, S.R., Legg, S.R.W., Cantle, J.P., Colantuoni, C., Carroll, J.B., & Ament, S.A. (2021). Single-nucleus RNA-seq reveals dysregulation of striatal cell identity due to Huntington's disease mutations. *Journal of Neuroscience*, *41*, 5534–5552.

Manjarrez, E., Curia, G., Stecina, K., & Lopez Valdes, A. (2023). Editorial: Bridging the gap between integrative neuroscience and translational neuroscience. *Frontiers in Integrative Neuroscience*, *17*, 1296701. https://doi.org/10.3389/fnint.2023.1296701

Mather, M., Cacioppo, J.T., & Kanwisher, N. (2013). Introduction to the special section: 20 years of fMRI-what has it done for understanding cognition? *Perspectives on Psychological Science: A Journal of the Association for Psychological Science*, *8*(1), 41–43. https://doi.org/10.1177/1745691612469036

Mathys, H., Davila-Velderrain, J., Peng, Z., Gao, F., Mohammadi, S., Young, J.Z., Menon, M., He, L., Abdurrob, F., Jiang, X., Martorell, A.J., Ransohoff, R.M., Hafler, B.P., Bennett, D.A., Kellis, M., & Tsai, L.H. (2019). Single-cell transcriptomic analysis of Alzheimer's disease. *Nature*, *570*, 332–337.

Memudu, A.E., Anzaku, F.A., Jibaniya, G.M., & Adanike, R.P. (2024). Neem leaf supplement ameliorates depressive like behaviour in Alzheimer's disease model in adult male Wistar rats. *Asian Journal of Research in Medical and Pharmaceutical Sciences*, *13*(1), 31–46. https://doi.org/10.9734/ajrimps/2024/v13i1245

Misrani, A., Tabassum, S., & Yang, L. (2021). Mitochondrial dysfunction and oxidative stress in Alzheimer's. *Frontiers in Aging Neuroscience*, *13*, 617588.

Mortada, I., Farah, R., Nabha, S., Ojcius, D.M., Fares, Y., Almawi, W.Y., & Sadier, N.S. (2021). Immunotherapies for neurodegenerative diseases. *Frontiers in Neurology*, *12*, 654739.

Naqvi, S., Panghal, A., & Flora, S.J.S. (2020). Nanotechnology: A promising approach for delivery of neuroprotective drugs. *Frontiers in Neuroscience*, *14*, 494. https://doi.org/10.3389/fnins.2020.00494

Ngoga, D., Mitchell, R., Kausar, J., Hodson, J., Harries, A., & Pall, H. (2014). Deep brain stimulation improves survival in severe Parkinson's disease. *Journal of Neurology, Neurosurgery, and Psychiatry*, *85*(1), 17–22. https://doi.org/10.1136/jnnp-2012-304715

Obrador, E., Salvador, R., López-Blanch, R., Jihad-Jebbar, A., Vallés, S.L., & Estrela, J.M. (2020). Oxidative stress, neuroinflammation and mitochondria in the pathophysiology of amyotrophic lateral sclerosis. *Antioxidants*, *9*, 901.

Onohuean, H., Agwu, E., & Nwodo, U.U. (2022). Systematic review and metaanalysis of environmental vibrio species—Antibiotic resistance. *Heliyon*, *8*, e08845. https://doi.org/10.1016/j.heliyon.2022.e08845

Onohuean, H., Akiyode, A.O., Akiyode, O., Igbinoba, S.I., & Alagbonsi, A.I. (2022) Epidemiology of neurodegenerative diseases in the East African region: A meta-analysis. *Frontiers in Neurology*, *13*, 1024004. https://doi.org/10.3389/fneur.2022.1024004

Pang, S.Y.-Y., Ho, P.W.-L., Liu, H.-F., Leung, C.-T., Li, L., Chang, E.E.S., Ramsden, D.B., & Ho, S.L. (2019). The interplay of aging, genetics and environmental factors in the pathogenesis of Parkinson's disease. *Translational Neurodegeneration*, *8*, 23.

Parmar, A., & Sarkar, S. (2016). Neuroimaging studies in obsessive compulsive disorder: A narrative review. *Indian Journal of Psychological Medicine*, *38*(5), 386–394. https://doi.org/10.4103/0253-7176.191395

Pathak, N., Vimal, S.K., Tandon, I., Agrawal, L., Hongyi, C., & Bhattacharyya, S. (2022). Neurodegenerative disorders of Alzheimer, Parkinsonism, amyotrophic lateral sclerosis and multiple sclerosis: An early diagnostic approach for precision treatment. *Metabolic Brain Disease*, *37*, 67–104. https://doi.org/10.1007/s11011-021-00800-w

Pickett, E.K., Herrmann, A.G., McQueen, J., Abt, K., Dando, O., Tulloch, J., Jain, P., Dunnett, S., Sohrabi, S., & Fjeldstad, M.P. (2019). Amyloid beta and tau cooperate to cause reversible behavioral and transcriptional deficits in a model of Alzheimer's disease. *Cell Reports*, *29*, 3592–3604.e5.

Popova, V., Daly, E.J., Trivedi, M., Cooper, K., Lane, R., Lim, P., Mazzucco, C., Hough, D., Thase, M.E., Shelton, R.C., Molero, P., Vieta, E., Bajbouj, M., Manji, H., Drevets, W.C., & Singh, J.B. (2019). Efficacy and safety of flexibly dosed esketamine nasal spray combined with a newly initiated Oral antidepressant in treatment-resistant depression: A randomized double-blind active-controlled study. *The American Journal of Psychiatry*, *176*(6), 428–438. https://doi.org/10.1176/appi.ajp.2019.19020172

Pozojevic, J., & Spielmann, M. (2023). Single-cell sequencing in neurodegenerative disorders. *Molecular Diagnosis and Therapy*, *27*, 553–561. https://doi.org/10.1007/s40291-023-00668-9

Pradhan, R., Singh, A.K., Kumar, P., Bajpai, S., Pathak, M., Chatterjee, P., Dwivedi, S., Dey, A.B. & Dey, S. (2022). Blood circulatory level of seven sirtuins in Alzheimer's disease: Potent biomarker based on translational research. *Molecular Neurobiology*, 1–12.

Rajanikant, G.K., & Kalthur, G. (2022). Growing importance of zebrafish in translational neuroscience. In *Zebrafish model for biomedical research*. 357–380, Springer, Singapore. https://doi.org/10.1007/978-981-16-5217-2_16

Rissardo, J.P., Vora, N.M., Tariq, I., Mujtaba, A., & Caprara, A.L.F. (2023). Deep brain stimulation for the management of refractory neurological disorders: A comprehensive review. *Medicina*, *59*(11), 1991. https://doi.org/10.3390/medicina59111991

Roberts, A.C. (2020). Prefrontal regulation of threat-elicited behaviors: A pathway to translation. *Annual Review of Psychology*, *71*, 357–387.

Salomone, S. (2020). New paradigms in neuroscience and related targets for drug discovery. *Frontiers in Pharmacology*, *11*, 750.

Santos, A.N., Kherif, F., Melie-Garcia, L., Lutti, A., Chiappini, A., Rauschenbach, L., Dinger, T.F., Riess, C., El Rahal, A., Darkwah Oppong, M., Sure, U., Dammann, P., & Draganski, B. (2023). Parkinson's disease may disrupt overlapping subthalamic nucleus and pallidal motor networks. *NeuroImage: Clinical*, *38*, 103432. https://doi.org/10.1016/j.nicl.2023.103432

Shahmoradian, S.H., Lewis, A.J., Genoud, C., Hench, J., Moors, T.E., Navarro, P.P., Castaño-Díez, D., Schweighauser, G., Graff-Meyer, A., Goldie, K.N., Sütterlin, R., Huisman, E., Ingrassia, A., Gier, Y., Rozemuller, A.J.M., Wang, J., Paepe, A., Erny, J., Staempfli, A., Hoernschemeyer, J., … Lauer, M.E. (2019). Lewy pathology in Parkinson's disease consists of crowded organelles and lipid membranes. *Nature Neuroscience*, *22*, 1099–1109.

Silveira, G. d. B., Muller, A.P., Machado-de-Ávila, R.A., & Silveira, P.C.L. (2021). Advance in the use of gold nanoparticles in the treatment of neurodegenerative diseases: New perspectives. *Neural Regeneration Research*, *16*(12), 2425–2426. https://doi.org/10.4103/1673-5374.313040

Smajić, S., Prada-Medina, C.A., Landoulsi, Z., Ghelfi, J., Delcambre, S., Dietrich, C., Jarazo, J., Henck, J., Balachandran, S., Pachchek, S., Morris, C.M., Antony, P., Timmermann, B., Sauer, S., Pereira, S.L., Schwamborn, J.C., May, P., Grünewald, A., & Spielmann, M. (2022). Single-cell sequencing of human midbrain reveals glial activation and a Parkinson-specific neuronal state. *Brain*, 145, 964–978.

Specht, K. (2020). Current challenges in translational and clinical fMRI and future directions. *Frontiers in Psychiatry*, *10*, 924.

Steinmetz, J.D., Seeher, K.M., Schiess, N., Nichols, E., Cao, B., Servili, C., Cavallera, V., Cousin, E., Hagins, H., Moberg, M.E. and Mehlman, M.L. (2024). Global, regional, and national burden of disorders affecting the nervous system, 1990–2021: A systematic analysis for the Global Burden of Disease Study 2021. *The Lancet Neurology*. *23*(4), 344–381 https://doi.org/10.1016/S1474-4422(24)00038-3

Stoyanov, D. S. (2017). Key developments in translational neuroscience: An update. *Balkan Medical Journal*, *34*(6), 485–486. https://doi.org/10.4274/balkanmedj.2017.6.0002

Strafella, C., Caputo, V., Galota, M.R., Zampatti, S., Marella, G., Mauriello, S., Cascella, R., & Giardina, E. (2018). Application of precision medicine in neurodegenerative diseases. *Frontiers in Neurology*, *9*, 701.

Sun, B.L., Li, W.W., Zhu, C., Jin, W.S., Zeng, F., Liu, Y.H., Bu, X.L., Zhu, J., Yao, X.Q., & Wang, Y.J. (2018). Clinical research on Alzheimer's disease: Progress and perspectives. *Neuroscience Bulletin*, *34*(6), 1111–1118. https://doi.org/10.1007/s12264-018-0249-z

Tanaka, M., Szabó, Á., & Vécsei, L. (2022). Integrating armchair, bench, and bedside research for behavioral neurology and neuropsychiatry. *Biomedicines*, *10*(12), 2999.

Tanaka, M., Szabó, Á, & Vécsei, L. (2023). Preclinical modeling in depression and anxiety: Current challenges and future research directions. *Advances in Clinical and Experimental Medicine*, *32*, 505–509.

Ubeda Matzilevich, E., Daniel, P.L., & Little, S. (2024). Towards therapeutic electrophysiological neurofeedback in Parkinson's disease. *Parkinsonism & Related Disorders*, *121*, 106010. https://doi.org/10.1016/j.parkreldis.2024.106010

Wallis, C.U., Cockcroft, G.J., Cardinal, R.N., Roberts, A.C., & Clarke, H.F. (2019). Hippocampal interaction with area 25, but not area 32, regulates marmoset approach–avoidance behavior. *Cerebral Cortex*, *29*(11), 4818–4830.

Wareham, L.K., Liddelow, S.A., Temple, S., Benowitz, L.I., Di Polo, A., Wellington, C., Goldberg, J.L., He, Z., Duan, X., Bu, G., Davis, A.A., Shekhar, K., Torre, A., Chan, D.C., Canto-Soler, M.V., Flanagan, J.G., Subramanian, P., Rossi, S., Brunner, T., Bovenkamp, D.E., … Calkins, D.J. (2022). Solving neurodegeneration: Common mechanisms and strategies for new treatments. *Molecular Neurodegeneration*, *17*, 23.

Woo, C.W., Chang, L.J., Lindquist, M.A., & Wager, T.D. (2017). Building better biomarkers: Brain models in translational neuroimaging. *Nature Neuroscience*, *20*(3), 365–377. https://doi.org/10.1038/nn.4478

Yang, W., Hamilton, J.L., Kopil, C., Beck, J.C., Tanner, C.M., Albin, R.L., Dorsey, E.R., Dahodwala, N., Cintina, I., Hogan, P., & Thompson, T. (2020). Current and projected future economic burden of Parkinson's disease in the U.S. *NPJ Parkinsons Disease*, *6*, 15. https://doi.org/10.1038/s41531-020-0117-1

Yousefi, M., Shadnoush, M., Sohrabvandi, S., Khorshidian, N., & Mortazavian, A.M. (2021). Encapsulation systems for delivery of flavonoids: A review. *Biointerface Research in Applied Chemistry*, *11*, 13934–13951.

Zeiss, C.J. (2017). From reproducibility to translation in neurodegenerative disease. *ILAR Journal*, *58*(1), 106–114.

Zhang, R., Liu, H., Pu, L., Zhao, T., Zhang, S., Han, K., & Han, L. (2023). Global burden of ischemic stroke in young adults in 204 countries and territories. *Neurology*, *100*(4), e422–e434. https://doi.org/10.1212/WNL.0000000000201467

Zheng, W., Cai, D.B., Xiang, Y.Q., Zheng, W., Jiang, W.L., Sim, K., Ungvari, G.S., Huang, X., Huang, X.X., Ning, Y.P., & Xiang, Y.T. (2020). Adjunctive intranasal esketamine for major depressive disorder: A systematic review of randomized double-blind controlled-placebo studies. *Journal of Affective Disorders*, *265*, 63–70. https://doi.org/10.1016/j.jad.2020.01.002

Zheng, J.C., & Chen, S. (2022). Translational neurodegeneration in the era of fast growing international brain research. *Translational Neurodegeneration*, *11*(1), 1–2.

2 Computational Models of Neurodegenerative Diseases

Akansha Gautam, Indranath Chatterjee, and Suruchi Gautam

2.1 NEURODEGENERATIVE DISEASES

Neurodegeneration refers to the gradual loss of neurons in the brain, starting with molecular dysregulation and culminating in significant regional dysfunction and eventual clinical disability. Neurodegenerative diseases (NDs) are a diverse group of disorders affecting the central nervous system (CNS), characterized by a chronic and selective degeneration of neuronal cells (Maselli et al., 2023). Numerous indications and symptoms that are associated with neurological disorders include memory problems, speech impairments, movement issues, cognitive and motor function challenges, and other coordination difficulties. These diseases are frequently identified by the buildup of particular proteins and structural weakness, showcasing various essential processes linked to advancing neuronal dysfunction and cell degeneration (Dugger & Dickson, 2017). Prominent examples of NDs include Alzheimer's disease (AD), multiple sclerosis (MS), Huntington's diseases, amyotrophic lateral sclerosis (ALS), Parkinson's disease, and Lewy body dementia (LBD). NDs affect over 40 million individuals globally (Wood et al., 2015). These conditions impact different levels of neuronal circuitry within the brain, from molecular alterations to systemic consequences. It leads to chronic conditions that progress gradually, with symptoms and effects manifesting later in life. It destroys components of the nervous system, particularly the brain.

Common features of NDs include progressive degeneration that involves the gradual loss of neurons and/or brain tissue over time, resulting in worsening symptoms and functional impairment. These aggregates can disrupt cellular function and contribute to neuronal dysfunction. Inflammatory processes within the brain and CNS play a role in the development of many NDs. Chronic inflammation can worsen neuronal damage and contribute to disease progression. These diseases may be brought on by environmental or hereditary factors. For example, diseases like Parkinson's disease and AD could arise from complex interactions between genetic susceptibility and environmental factors. These diseases display significant diversity in their clinical presentation, underlying pathology, and disease progression. This diversity presents obstacles for the diagnosis, treatment, and research of these conditions.

One of the most common NDs is AD. This condition is the leading cause of dementia, impacting memory, cognitive abilities, and behavior. It is identified by the buildup of abnormal protein deposits in the brain, including amyloid plaques and tau tangles. The impairment of brain activity and cognitive functions in AD and other related disorders is often linked to the degeneration of specific brain regions which are basically a large group of cells responsible for brain functioning. However, the relationship between neurodegenerative processes, functional systems, and the clinical traits of dementia remains unclear (Jones et al., 2022). Symptoms initially present as mild memory impairments, which can sometimes be mistaken for age-related memory decline. While lifestyle factors may not directly impact the pathology of AD, they can still play a significant role in improving

DOI: 10.1201/9781032711102-2

outcomes for individuals living with the condition (Scheltens et al., 2021). As the condition progresses, individuals may experience severe memory impairments, personality alterations, speech related problems, motor impairments, hallucinations, and delusions (Joshi et al., 2010). The diagnostic criteria for AD involve the identification of AD biomarkers through magnetic resonance imaging (MRI) or positron emission tomography (PET) imaging, as well as an evaluation of symptoms of dementia and the level of intellectual disability (Huntington Study Group Reach2HD Investigators, 2015).

Parkinson's disease: Parkinson's disease is the second most common ND, characterized by symptoms typical of hypokinetic disorders. It is often considered a representative example of such conditions (Tăuţan et al., 2021). Parkinson's primarily affects movement and is caused by the reduction in dopamine-producing brain cells. This leads to symptoms like tremors, stiffness, sleep disturbances, small handwriting, and difficulties with balance and coordination. Approximately 3%–5% of Parkinson's disease cases are attributed to genetic mutations in known Parkinson's-related genes, resulting in what is termed monogenic Parkinson's disease. Conversely, 16%–36% of the risk of non-monogenic Parkinson's disease is attributed to a set of 90 genetic risk variants (Bloem et al., 2021). A cure for the disease has not yet been found, so current treatments primarily aim to alleviate symptoms. This can be achieved through deep brain stimulation, medication, or physical therapy. Understanding the progression of visible changes in individuals with an increased risk of developing dementia in Parkinson's disease could serve as a comprehensive biomarker for the identification and management of those who are likely to develop Parkinson's dementia.

Huntington's disease: Huntington's disease is a genetic disorder that is characterized by progressive movement and cognitive and psychiatric symptoms. This condition is caused by a mutation in the huntingtin (HTT) gene, which results in the production of abnormal HTT protein that damages nerve cells in the brain. At the cellular level, mutant HTT leads to neuronal dysfunction through various mechanisms, including disruption of proteostasis, transcription, and mitochondrial function, as well as direct toxicity of the mutant protein which results in impaired cellular processes and ultimately contributes to the pathogenesis of Huntington's disease (McColgan & Tabrizi, 2018). The prevalence of this condition in the Caucasian population is estimated to be between 1 in 10,000 and 1 in 20,000. The average age at which symptoms typically begin is between 30 and 50 years old (Roos, 2010). The symptoms of this condition include chorea, which is characterized by uncontrolled movements, abnormal body postures, speech impairments, and changes in behavior, emotion, judgment, and cognition. Diagnosis is typically confirmed through genetic testing and neuroimaging techniques.

ALS is a progressive ND that affects motor neurons in the brain and spinal cord. This degeneration of motor neurons results in the gradual loss of voluntary muscle control, ultimately leading to muscle weakness, paralysis, and, in severe cases, respiratory failure. Additionally, some individuals with ALS may experience cognitive and behavioral changes. The incidence of sporadic ALS is approximately 1.89 per 100,000 people per year, while the prevalence is around 5.2 per 100,000 individuals as these rates remain relatively consistent across Western countries, with some areas in the Western Pacific experiencing higher frequencies. The average age of the onset of sporadic ALS is typically around 60 years old (Wijesekera & Nigel Leigh, 2009). ALS is characterized by the atrophy of muscles due to the loss of control over them. The mechanisms responsible for the development of ALS are not well understood in most patients. However, a subset of patients have familial ALS and carry mutations in genes that are crucial for neuronal function as these genetic mutations can significantly impact the progression and severity of the disease (Hardiman et al., 2017). The initial stages of the disease typically manifest in the limbs and later progress swiftly to other areas of the body. In advanced stages of ALS, the muscles responsible for managing the respiratory system start to debilitate, posing a significant threat to the individual's ability to breathe properly.

Multiple system atrophy (MSA) is a rare and progressive neurodegenerative disorder that impacts various systems within the CNS, such as the autonomic nervous system (ANS) and motor systems. MS is a chronic disease that manifests differently in each individual. While some may experience a mild form of the disease with minimal disability, others may face a progressive deterioration of their condition, resulting in increased disability over time. The estimated population prevalence of MSA is 4 per 100,000, with wide confidence intervals, and Parkinson's disease is the most common cause of misdiagnoses of MSA (Quinn, 2004). It is a type of atypical parkinsonism that has an unknown cause. There are two primary subtypes of MSA: MSA with predominant parkinsonism (MSA-P) and MSA with predominant cerebellar ataxia (MSA-C). These subtypes are differentiated by their primary clinical symptoms. Its symptoms necessitate a multifaceted approach to address autonomic failure, alleviate parkinsonism, cerebellar ataxia, and related disabilities (Poewe et al., 2022).

NDs pose significant challenges as their etiology remains elusive, and a definitive cure has yet to be found. Many of these diseases do not have treatments to slow or stop their progression, largely because it is difficult to unravel the complex and simultaneous presentation of various pathophysiological changes. All of these diseases have debilitating conditions that significantly impact an individual's quality of life and require comprehensive medical management to address its various symptoms and complications. However, current technologies and computational capabilities offer hope for untangling these changes and discovering new mechanisms that drive the onset and progression of NDs.

2.2 MEDICAL IMAGING SYSTEM ROLE IN NEURODEGENERATIVE DISEASES

Medical science, also referred to as biomedical science or health science, is a crucial branch of science that focuses on the health of patients and healing processes. It is a diverse field of scientific disciplines dedicated to comprehending the human body, diseases, and medical treatments. It incorporates an extensive range of research, study, and practical applications aimed at enhancing our understanding of human health and enhancing medical care. The rapid advancements in the field of medical science have greatly benefited society as a whole. Medical science involves the in-depth study of how the human body functions when in good health. The human body is incredibly complex in its structure, requiring significant effort to maintain proper function throughout a lifetime. Healthcare practitioners play a crucial role in caring for patients, overseeing the diagnosis of illnesses, and providing treatments to alleviate the severity of diseases.

Humans are naturally inclined to learn, a feature that, in today's culture, has become essential to machines (Shinde & Shah, 2018). A medical imaging system is an advanced technology utilized to generate visual representations of the interior of the human body for medical diagnosis and care. These systems enable healthcare professionals to visualize anatomical structures, identify abnormalities, monitor physiological processes, and guide interventions. Medical visuals are crucial in making informed decisions promptly. Producing a variety of images is easily achievable using various techniques of image processing as these techniques can be applicable to medical images in order to understand them meticulously and accurately (Azhari et al., 2014). Different technologies provide specific information about various areas of the human body, aiding in the diagnosis and treatment of diseases, injuries, and illnesses. This technique significantly enhances the accuracy of results in modern medicine, offering an advantage at every stage of its application. It is an indispensable tool for improving patient care, conducting thorough clinical analyses, and facilitating better medical decision-making through visual representations.

Medical imaging is the process of using various technologies to create images of the interior organs or tissues of the body, which are used by healthcare professionals to observe, diagnose, and treat medical conditions (Gautam & Chatterjee, 2023). A medical imaging system consists of several key components that work together to generate, process, display, store, and integrate imaging data. These components include (1) Image Acquisition Device: Using this component, images are created by the emission or detection of different forms of energy, including X-rays, gamma rays, sound waves, and radio waves. (2) Data Processing Unit: The data processing unit is responsible for processing raw image data that has been collected by the acquisition device. It then reconstructs this data into a format that is easily interpretable. (3) Display Device: The display device is responsible for presenting the processed images to healthcare providers for analysis and interpretation. In modern systems, high-resolution monitors are commonly used for image display, allowing for clear and detailed visualization of medical images. (4) Storage and Archiving System: Medical imaging systems are equipped with storage and archiving capabilities to securely store images and allow for easy access for future reference, comparison, and analysis. This feature ensures that medical professionals can efficiently retrieve and analyze images when needed, ultimately improving patient care and treatment outcomes. (5) Integration with Healthcare Information Systems: Connecting with electronic health records (EHRs) and other healthcare information systems enables effortless sharing, retrieval, and management of imaging data throughout the healthcare enterprise. This seamless integration streamlines processes and enhances efficiency within the healthcare system.

Medical imaging plays a crucial role in the diagnosis and monitoring of NDs. One of the most commonly used techniques is MRI, which provides detailed images of the brain's structure. MRI is highly effective in detecting structural abnormalities in the brain, such as atrophy (shrinkage) of specific regions that are linked to disease progression. For instance, in AD, MRI can reveal hippocampal atrophy, a key indicator of memory impairment. Computed tomography (CT) scans offer precise cross-sectional images of the brain. Although not as frequently utilized as MRI, CT scans can be used to identify structural abnormalities, such as brain atrophy or the presence of lesions, in NDs. PET imaging enables the visualization of metabolic processes in the brain by detecting the distribution of radiotracers that are selectively absorbed by active brain regions. This technology is particularly useful in assessing changes in glucose metabolism or the accumulation of abnormal proteins associated with NDs. Functional magnetic resonance imaging (fMRI) is a highly advanced imaging method that can perceive changes in oxygen levels in order to create detailed maps of brain activity. Although fMRI is commonly utilized in research settings, it also offers valuable insights into functional alterations in the brain linked to NDs. fMRI is a crucial tool in the development of automated methods for detecting schizophrenia (Chatterjee et al., 2020). These changes may include shifts in brain connectivity or activity patterns. Physicians utilize a variety of imaging modalities, including MRI, ultrasound, PET, CT scans, and X-ray, to detect, diagnose, and treat serious illnesses (Serte et al., 2020). These advanced techniques are essential in the medical field as they provide in-depth insights into the human body, allowing healthcare professionals to make precise and timely diagnosis.

2.3 COMPUTATIONAL MODELS OF NEURODEGENERATIVE DISEASES

After analyzing the medical outlook on neurodegeneration, identifying the eminent diseases associated with it, our focus shifts to the current computational approaches that aid in diagnosing, enhancing, and monitoring the quality of life for patients. Computational models are essential in the study of NDs as they help in unraveling the underlying mechanisms, pinpointing potential therapeutic targets, and advancing the development of innovative treatments. These models vary from basic mathematical equations to intricate simulations that incorporate neural networks and computational biology techniques. Computational models provide vital tools for understanding NDs from various perspectives, integrating experimental data, and expediting the development of effective therapies. These models combine data from a variety of sources, such as clinical observations, neuroimaging, molecular biology, and genetics to offer valuable insights into the pathophysiology of diseases and to help steer the development of potential therapies. The use of computational methods in assisting patients with neurodegenerative disorders serves multiple purposes. These methods can aid in diagnosis by accurately characterizing symptoms and pinpointing the disease precisely. Furthermore, they can be utilized to establish an objective monitoring system to track the progression of diseases. By merging computational modeling with experimental research, scientists can acquire a thorough understanding of the fundamental processes of these debilitating disorders, as a goal to enhance patients' outcomes.

Creating a generic computational model involves several key steps. First, data acquisition is essential, which entails gathering necessary data from various sources. Preprocessing is then performed to address issues like missing values, noise, outliers, and data normalization. Next, feature extraction and selection techniques are utilized to identify and choose relevant information from raw data. It is crucial to select an appropriate computational modeling approach based on the research question and available data. Following this, mathematical, statistical, or computational models are developed to represent the underlying biological processes or phenomena related to NDs. The trained models' performance is evaluated using a validation set or, ideally, an independent test set. Ultimately, these steps culminate in the creation of a black-box computational tool used to identify NDs and draw biological interpretations from the data (Figure 2.1).

There are several common types of computational models used in the research of NDs (See Table 2.1). These include mathematical models, network models, machine learning models, agent-based models (ABMs), and biophysical models.

Mathematical models are quantitative frameworks that utilize mathematical equations to describe the behavior of biological systems implicated in NDs. These models provide a structured approach to understanding the complex interactions within the body that support the emergence and advancement of these situations. Simulation of different aspects of disease progression can be performed using these models, including protein aggregation, cognitive decline, neuronal loss, and synaptic dysfunction. Such models simulate the kinetics of protein aggregation, providing insight into the processes involved in the formation of toxic protein aggregates, which are features of NDs like Alzheimer's and Parkinson's. Bukkuri (2019) has proposed a mathematical model that accurately represents the biochemical interactions between ascorbic acid (AA), dehydroascorbic acid (DHA), and reactive oxygen

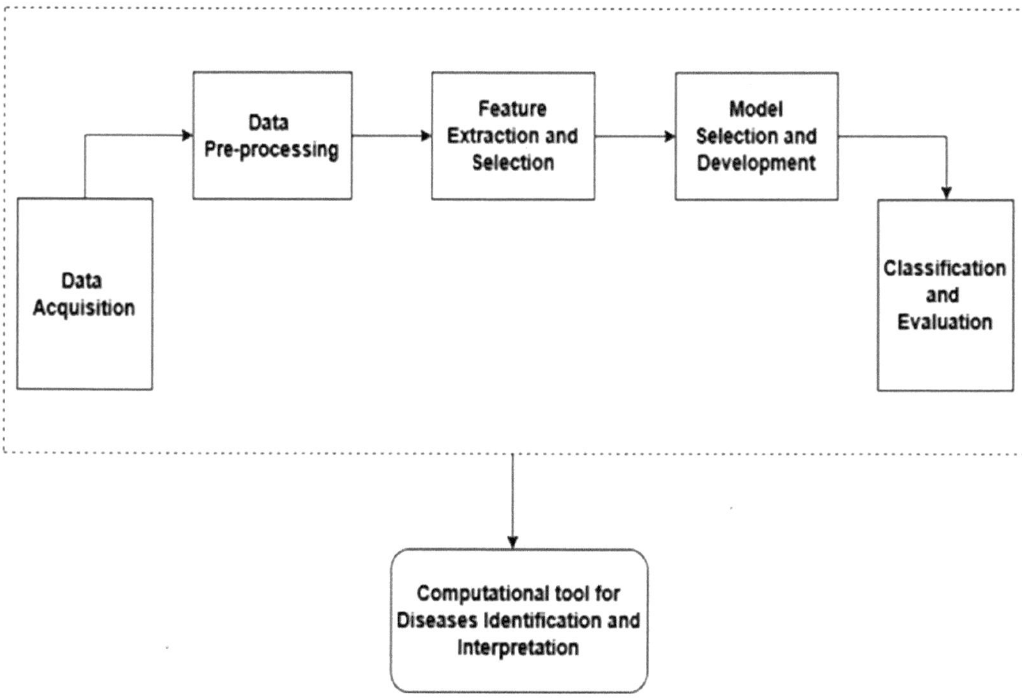

FIGURE 2.1 Steps involved in computational model.

species (ROS) in the brain. This model has been utilized to conduct simulations under both normal physiological conditions and in situations of NDs. By employing mathematical models, we are able to comprehend the dynamics on a deeper level of diseases and predict the potential impact of interventions on disease outcomes. These models are utilized to analyze data in order to quantify biomarkers of disease pathology. They establish correlations between imaging biomarkers and clinical measures, ultimately enhancing disease diagnosis, prognosis, and treatment monitoring.

Biophysical models are utilized to simulate the physical properties of neurons, synaptic transmission, ion channels, and encompassing electrical activity. These models play a crucial role in investigating how alterations at the cellular level impact NDs like Alzheimer's and Parkinson's. Such models are used to study the interactions between pathological proteins and cellular components, such as membranes, cytoskeletal proteins, and synaptic proteins. They aim to understand how these interactions disrupt cellular function and lead to neuronal degeneration. Iturria-Medina et al. (2014) have developed a stochastic epidemic spreading model (ESM) as a biophysical model of misfolded proteins (MP) propagation and deposition. This model allows for the reconstruction of each person's lifetime MP propagation history within the brain, as well as ensuing factor analysis that facilitates propagation and deposition. Through the integration of experimental data with computational simulations, researchers are able to obtain valuable insights into the mechanisms that drive disease pathology (Figure 2.2).

Different machine learning models, including deep learning, in NDs are widely used for analyzing vast datasets associated with NDs. These advanced techniques have become invaluable tools for enhancing the diagnosis and disease monitoring processes

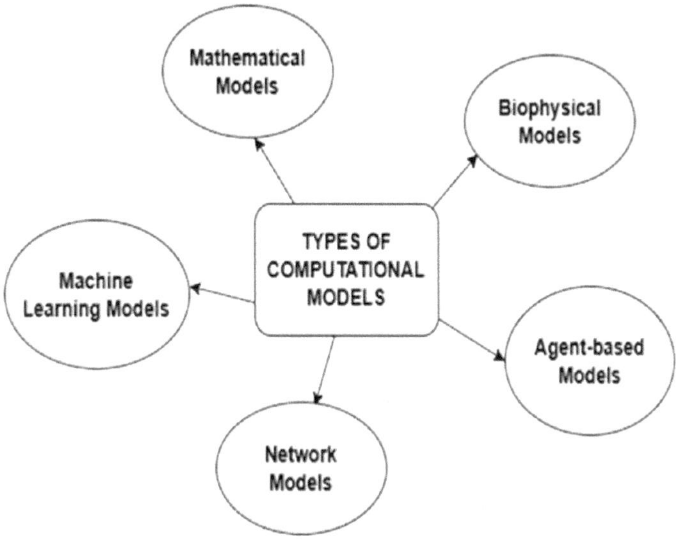

FIGURE 2.2 Different types of computational models used in neurodegenerative diseases.

(Tăuţan et al., 2021). They play a crucial role in identifying disease biomarkers, forecasting disease progression, and categorizing patients according to their treatment response. Machine learning is a prominent field in modern computing, while deep learning is a subfield of machine learning focused on mimicking the human brain. Machine learning is a subset of artificial intelligence (AI) that allows systems to learn from data without the need for explicit programming as this technology enables machines to classify data, make predictions, and identify trends without the constant need for human intervention (Kim et al., 2019). Deep learning is inspired by the neurons in the human brain and aims to replicate their functionality. These models are utilized to classify patients based on clinical features, genetic markers, and imaging biomarkers, aiding in the early diagnosis and differential diagnosis of NDs. By utilizing machine learning, machines are able to make data-driven decisions to successfully complete tasks. Moreover, machine learning methodologies aid in the exploration of new drug candidates and the repurposing of current drugs for the treatment of NDs. Recent advancements in AI have opened up new possibilities for improving diagnosis and monitoring in healthcare. Computational approaches, such as detecting disease onset, characterizing diseases, enhancing differential diagnosis, quantifying disease progression, and tracking medication effects, can all benefit from these technologies. Machine learning algorithms can automate and enhance these tasks, making them more efficient and accurate. One common thread among various machine learning applications is the incorporation of multiple high-dimensional data sources, each offering a unique perspective on diseases. This integration allows for the automated extraction of valuable insights that can be put into action (Myszczynska et al., 2020).

Network models are representations of the intricate connectivity patterns found within neural circuits in the brain. It is centered on comprehending how neuronal networks and brain connectivity are impacted by pathological processes, including protein aggregation, synaptic dysfunction, and neuronal loss. By simulating the effects of NDs on these networks, we can gain insight into how information processing is impacted. These models provide valuable insights into how alterations in network dynamics play a role in cognitive decline, motor symptoms, uncover disease mechanisms, and forecast disease progression and other disease manifestations. Network models utilize graph theory to depict the brain as a network consisting of nodes representing brain regions which are interconnected by edges that show structural or functional connections between different brain regions. Graph measures, including degree centrality, clustering coefficient, and betweenness centrality, are employed to measure the characteristics of networks and organization. Network models also leverage connectomics data to demonstrate the brain's anatomical and functional connections. Connectomics involves mapping the connections between different brain regions using techniques like diffusion MRI, tractography, and resting-state fMRI. Vogel et al. (2023) delved into the cutting-edge field of neuroimaging-based connectomics, exploring its potential for mapping and predicting neurodegenerative processes.

ABMs imitate how individual agents—like immune cells and neurons—behave when they interact with one another in a spatial environment, for example, ABM replicate the spatial organization of the brain by depicting brain regions, neuronal networks, and anatomical connections. This simulation allows for a detailed examination of how these components interact and influence brain function. These models are able to depict the complex interplay between different cell types as well as the heterogeneous nature of neurodegenerative disorders. They are especially useful for studying processes in the brain like inflammation, neuroinvasion, and the spread of disease-causing proteins. ABMs are becoming more prevalent in the modeling of infectious disease transmission; however, there is a lack of focus in the literature on the implementation and performance of these models (Willem et al., 2015).

TABLE 2.1
Related Studies on Neurodegenerative Diseases Using Different Computational Models

Author	Year	Type of Computational Model	Applied Algorithm/Model
Kepp	2019	Mathematical model	The model of proteomic cost minimization
Trambaiolli et al.	2017	Machine Learning model	Support vector machine for classification of the disease
Khanal et al.	2016	Biophysical model	Biomechanics-based tissue loss model
Manzoni et al.	2020	Network model	Graph theory
Abdelgawad et al.	2023	Agent-based model	Susceptible-infected-removed (SIR) model
Ficiarà et al.	2020	Mathematical model	A mathematical model for evaluation of iron transport
Lin et al.	2020	Machine learning model	Multiplex blood biomarkers based-machine learning model
Iturria-Medina et al.	2014	Biophysical model	Epidemic spreading model to characterize the propagation of misfolded proteins
Santiago & Potashkin	2014	Network model	Network analysis for identifying biologically meaningful biomarkers
Pennisi et al.	2013	Agent-based model	An agent-based model has been used to accurately replicate the characteristic oscillatory behavior commonly seen in the most prevalent form of multiple sclerosis

2.4 APPLICATIONS OF COMPUTATIONAL MODELS OF NEURODEGENERATIVE DISEASES

A wide range of conditions known as neurodegenerative illnesses are defined by the progressive degeneration of the central or peripheral nervous systems. These conditions result in a gradual decline in cognitive function, motor skills, and overall quality of life. We are in dire need of effective treatments, which can only be developed through a thorough understanding of the causes and mechanisms underlying each disease. Computational models of NDs have a wide range of applications in different fields such as basic research, drug discovery, clinical decision-making, and personalized medicine. These models can help in understanding mechanisms behind the diseases. One exciting application of computational methods is in predicting disease events and even identifying and classifying diseases before clinical symptoms appear. In the case of NDs, the process can begin years before any symptoms become apparent. Computational tools can also be applied in neuroimaging analysis, where computational models are utilized to analyze and interpret the data retrieved through neuroimaging techniques like MRI, fMRI, and PET. Another essential application of computational models could be in the field of drug discovery as they aid in the identification and optimization of potential therapeutic compounds. Utilizing computational models can provide valuable insights into predicting the advancement of NDs by analyzing clinical and biological data collected from patients. Computational models of NDs are essential in enhancing our comprehension of these intricate disorders, expediting drug discovery initiatives, enhancing patient care, and ultimately, propelling the creation of successful treatments to mitigate the effects of these debilitating conditions. These models are essential in advancing our knowledge of NDs and in the successful development of treatments.

2.5 CHALLENGES IN COMPUTATIONAL MODELING

Computational approaches have the potential to greatly enhance our understanding of NDs and aid in the development of effective treatments. However, they also face numerous challenges. These challenges stem from the complex nature of diseases as they possess intricate conditions that arise from a blend of genetic, environmental, and lifestyle factors. Grasping the complete spectrum of disease mechanisms and their interactions in computational models presents a significant challenge in the field. Limitations in data availability and accuracy could be another legitimate challenge. Computational modeling is heavily dependent on a diverse range of data sources, including clinical data, imaging data, molecular data, and longitudinal datasets. Nevertheless, the precision and dependability of these models may be jeopardized by data quality concerns such as missing values, measurement errors, and variations among different sources. Expertise and problem-solving abilities are essential in this field and the computational methods employed. It requires a multidisciplinary approach that integrates expertise from various fields such as neuroscience, genetics, bioinformatics, computational biology, and clinical medicine. Through collaboration across disciplines, diverse perspectives, knowledge, and skills can be leveraged to tackle complex issues and develop state-of-the-art computational models. It requires effective communication and cooperation among experts from various domains as a crucial step for identifying pertinent research questions, developing suitable methodologies, and interpreting computational findings within the realm of clinical significance. In today's rapidly evolving landscape, it is imperative to remain at the cutting edge and embrace new technologies in order to achieve success. By directly addressing these challenges and continuously enhancing our techniques, we can unlock the full potential of computational modeling and drive innovation in the field of healthcare.

2.6 CONCLUSION AND SCOPE OF FUTURE WORK

In recent years, there has been a notable surge in research on NDs through the utilization of computational methods. This trend reflects a growing recognition of the potential benefits of integrating technology and data analysis in the study of these complex conditions. The growing economic burden caused by these diseases has created a demand for quicker results and innovative solutions in automatic diagnostic and health assessment systems. Numerous proposed solutions involve the integration of various models and algorithms. NDs are characterized by permanent and currently incurable conditions. However, advancements in medical science have made many of these diseases treatable. The primary goal now is to effectively manage symptoms and slow the progression of these disorders. This goal can be accomplished by utilizing computational methods to promptly monitor and diagnose neurodegenerative diseases.

REFERENCES

Abdelgawad, A., Rahayel, S., Zheng, Y. Q., Tremblay, C., Vo, A., Misic, B., & Dagher, A. (2023). Predicting longitudinal brain atrophy in Parkinson's disease using a susceptible-infected-removed agent-based model. *Network Neuroscience*, *7*(3), 906–925.

Azhari, E. E. M., Hatta, M. M. M., Htike, Z. Z., & Win, S. L. (2014). Tumor detection in medical imaging: A survey. *International Journal of Advanced Information Technology*, *4*(1), 21.

Bloem, B. R., Okun, M. S., & Klein, C. (2021). Parkinson's disease. *The Lancet*, *397*(10291), 2284–2303.

Bukkuri, A. (2019). A mathematical model of the effects of ascorbic acid on the onset of neurodegenerative diseases. *Open Journal of Mathematical Sciences*, *3*(1), 300–309.

Chatterjee, I., Kumar, V., Rana, B., Agarwal, M., & Kumar, N. (2020). Impact of ageing on the brain regions of the schizophrenia patients: An fMRI study using evolutionary approach. *Multimedia Tools and Applications*, *79*(33), 24757–24779.

Dugger, B. N., & Dickson, D. W. (2017). Pathology of neurodegenerative diseases. *Cold Spring Harbor Perspectives in Biology*, *9*(7), a028035.

Ficiarà, E., D'Agata, F., Ansari, S., Boschi, S., Rainero, I., Priano, L., Cattaldo, S., Abollin, O., Cavalli, R., & Guiot, C. (2020). A mathematical model for the evaluation of iron transport across the blood-cerebrospinal fluid barrier in neurodegenerative diseases. *2020 42nd annual international conference of the IEEE Engineering in Medicine & Biology Society (EMBC)* (pp. 2270–2273). IEEE.

Gautam, A., & Chatterjee, I. (2023). Medical imaging and schizophrenia: A study on state-of-art applications. *Cognizance of schizophrenia: A profound insight into the psyche* (pp. 271–281). Springer Nature Singapore.

Hardiman, O., Al-Chalabi, A., Chio, A., Corr, E. M., Logroscino, G., Robberecht, W., Shaw, P., Simmons, Z., & Van Den Berg, L. H. (2017). Amyotrophic lateral sclerosis. *Nature Reviews Disease Primers*, *3*(1), 1–19.

Huntington Study Group Reach2HD Investigators. (2015). Safety, tolerability, and efficacy of PBT2 in Huntington's disease: A phase 2, randomised, double-blind, placebo-controlled trial. *The Lancet Neurology*, *14*(1), 39–47.

Iturria-Medina, Y., Sotero, R. C., Toussaint, P. J., Evans, A. C., & Alzheimer's Disease Neuroimaging Initiative. (2014). Epidemic spreading model to characterize misfolded proteins propagation in aging and associated neurodegenerative disorders. *PLoS Computational Biology*, *10*(11), e1003956.

Jones, D., Lowe, V., Graff-Radford, J., Botha, H., Barnard, L., Wiepert, D., Murphy, M. C., Murray, M., Senjem, M., Gunter, J., Wiste, H., Boeve, B., Knopman, D., Petersen, R., & Jack, C. (2022). A computational model of neurodegeneration in Alzheimer's disease. *Nature Communications*, *13*(1), 1643.

Joshi, S., Shenoy, P. D., Venugopal, K. R., & Patnaik, L. M. (2010). Classification of neurodegenerative disorders based on major risk factors employing machine learning techniques. *International Journal of Engineering and Technology*, 2(4), 350.

Kepp, K. P. (2019). A quantitative model of human neurodegenerative diseases involving protein aggregation. *Neurobiology of Aging*, *80*, 46–55.

Khanal, B., Lorenzi, M., Ayache, N., & Pennec, X. (2016). A biophysical model of brain deformation to simulate and analyze longitudinal MRIs of patients with Alzheimer's disease. *NeuroImage*, *134*, 35–52.

Kim, M., Yun, J., Cho, Y., Shin, K., Jang, R., Bae, H. J., & Kim, N. (2019). Deep learning in medical imaging. *Neurospine*, *16*(4), 657.

Lin, C. H., Chiu, S. I., Chen, T. F., Jang, J. S. R., & Chiu, M. J. (2020). Classifications of neurodegenerative disorders using a multiplex blood biomarkers-based machine learning model. *International Journal of Molecular Sciences*, *21*(18), 6914.

Manzoni, C., Lewis, P. A., & Ferrari, R. (2020). Network analysis for complex neurodegenerative diseases. *Current Genetic Medicine Reports*, *8*, 17–25.

Maselli, F., D'Antona, S., Utichi, M., Arnaudi, M., Castiglioni, I., Porro, D., Papaleo, E., Gandellini, P., & Cava, C. (2023). Computational analysis of five neurodegenerative diseases reveals shared and specific genetic loci. *Computational and Structural Biotechnology Journal*, *21*, 5395–5407.

McColgan, P., & Tabrizi, S. J. (2018). Huntington's disease: A clinical review. *European Journal of Neurology*, *25*(1), 24–34

Myszczynska, M. A., Ojamies, P. N., Lacoste, A. M., Neil, D., Saffari, A., Mead, R., Hautbergue, G. M., Holbrook, J. D., & Ferraiuolo, L. (2020). Applications of machine learning to diagnosis and treatment of neurodegenerative diseases. *Nature Reviews Neurology*, *16*(8), 440–456.

Pennisi, M., Rajput, A. M., Toldo, L., & Pappalardo, F. (2013). Agent based modeling of Treg-Teff cross regulation in relapsing-remitting multiple sclerosis. *BMC Bioinformatics*, *14*, 1–14.

Poewe, W., Stankovic, I., Halliday, G., Meissner, W. G., Wenning, G. K., Pellecchia, M. T., Seppi, K., Palma, J. A., & Kaufmann, H. (2022). Multiple system atrophy. *Nature Reviews Disease Primers*, *8*(1), 56.

Quinn, N. (2004). Multiple system atrophy. *Primer on the autonomic nervous system* (pp. 290–292). Academic Press.

Roos, R. A. (2010). Huntington's disease: A clinical review. *Orphanet Journal of Rare Diseases*, *5*, 1–8.

Santiago, J. A., & Potashkin, J. A. (2014). A network approach to clinical intervention in neurodegenerative diseases. *Trends in Molecular Medicine*, *20*(12), 694–703.

Scheltens, P., De Strooper, B., Kivipelto, M., Holstege, H., Chételat, G., Teunissen, C. E., Cummings, J., & van der Flier, W. M. (2021). Alzheimer's disease. *The Lancet*, *397*(10284), 1577–1590.

Serte, S., Serener, A., & Al-Turjman, F. (2020). Deep learning in medical imaging: A brief review. *Transactions on Emerging Telecommunications Technologies*, *33*(10), e4080.

Shinde, P. P., & Shah, S. (2018). A review of machine learning and deep learning applications. *2018 Fourth international conference on computing communication control and automation (ICCUBEA)* (pp. 1–6). IEEE.

Tăuţan, A. M., Ionescu, B., & Santarnecchi, E. (2021). Artificial intelligence in neurodegenerative diseases: A review of available tools with a focus on machine learning techniques. *Artificial Intelligence in Medicine*, *117*, 102081.

Trambaiolli, L. R., Spolaôr, N., Lorena, A. C., Anghinah, R., & Sato, J. R. (2017). Feature selection before EEG classification supports the diagnosis of Alzheimer's disease. *Clinical Neurophysiology*, *128*(10), 2058–2067.

Vogel, J. W., Corriveau-Lecavalier, N., Franzmeier, N., Pereira, J. B., Brown, J. A., Maass, A., Botha, H., Seeley, W. W., Bassett, D. S., Jones, D. T., & Ewers, M. (2023). Connectome-based modelling of neurodegenerative diseases: Towards precision medicine and mechanistic insight. *Nature Reviews Neuroscience*, *24*(10), 620–639.

Wijesekera, L. C., & Nigel Leigh, P. (2009). Amyotrophic lateral sclerosis. *Orphanet Journal of Rare Diseases*, *4*(1), 22.

Willem, L., Stijven, S., Tijskens, E., Beutels, P., Hens, N., & Broeckhove, J. (2015). Optimizing agent-based transmission models for infectious diseases. *BMC Bioinformatics*, *16*, 1–10.

Wood, L. B., Winslow, A. R., & Strasser, S. D. (2015). Systems biology of neurodegenerative diseases. *Integrative Biology*, *7*(7), 758–775.

3 A Survey on Machine Learning Applications in Neuroimaging

Himani Sharma, Aditya Azad Singh, Aman Pratap Singh, and Ashish

3.1 INTRODUCTION

Neuroimaging is the process of acquiring images of the brain and nervous system via various techniques such as magnetic resonance imaging (MRI), functional magnetic resonance imaging (fMRI), and positron emission tomography (PET). On the other hand, neuroimaging analysis is employed to process and interpret these images to understand how the brain works, assess brain health, and diagnose/detect diseases such as tumors or strokes.

In recent years, machine learning (O'Donnell & Schultz, 2015; Grueso and Viejo-Sobera, 2021) and deep learning methodologies (Quaak et al., 2021; Chatterjee et al., 2023; Zhang et al., 2020) have put a profound impact on the field of neuroimaging analysis for understanding the complexities of human brain and impairments associated. These approaches are capable of decoding how neurological disorders affect the human body. Machine learning is a branch of artificial intelligence (AI) which allows computers to utilize the neuroimages to extract patterns or trends which are used to predict possible neurological disorders in individuals. Moreover, the speed of machine learning-based analyzers is far superior to that of their human counterparts which offers a great advantage in time-sensitive cases like strokes. Furthermore, ensemble learning and deep learning techniques are being increasingly used (Kumar et al., 2024). Deep learning makes use of various neurological features arranged into many deep neural network layers. These layers are later used in accurately predicting neurological disorders. It outshines machine learning in identifying trends and anomalies in computed tomography (CT) scans, MRIs, fMRIs, etc.

Mateos-Pérez et al. (2018) discuss various machine learning techniques applied on MRI data to make a prediction about various brain disorders. To remove the irrelevant features and cleaning from our neurological data, various dimensionality reduction techniques are used like linear discriminant analysis (LDA). Preprocessing of data is done to make data suitable for machine learning analysis. Various machine learning algorithms are used to analyze neuroimages, and each has its own advantages and disadvantages.

Support vector machine (SVM) is a well-known supervised learning algorithm which is used to classify various types of neurological disorders. Supervised algorithms are capable of generalizing rules or patterns from a labeled set of input data and classify accordingly. These are used to decrypt patterns that help us classify various diseases like tumors and blockage based on features extracted from MRI scans. It is known for its robustness against overfitting. Li et al. (2024) utilized Support Vector Machines (SVM) to classify patients with mild cognitive impairment (MCI) a transitional stage in the elderly that can potentially progress to Alzheimer's disease (AD). Xiao et al. (2024) have shown SVM's good performance in classifying Parkinson's disease.

DOI: 10.1201/9781032711102-3

k-Nearest neighbors (KNNs) is a simple supervised learning algorithm which makes prediction based on the class label of the neighboring data points. It is frequently used to predict cognitive decline from fMRI data. Tuncer et al. (2024) used KNN along with SVM to propose a self-organized feature model to identify specific language impairment in children with Parkinson's disease.

Decision trees (DTs) can effectively predict peculiar regions of brain affected by neurological disorders such as schizophrenia (Chatterjee et al., 2019). It can handle both categorical and numeric data. Shahid and Iftikhar (2022) have applied classification tree to identify neuropsychiatric problems in various brain tumors.

Ensemble learning is an approach to combine various machine learning models sequentially or parallelly. Random forest (RF) is an ensemble learning technique that utilizes multiple DTs to predict an output. Based on features extracted from PET scans, Sarica et al. (2017) were able to classify the stages of AD. Also, it can handle missing data and outliers efficiently. Nour et al. (2023) used ensemble learning approach and dynamic classifier selection approach to classify Parkinson's disease using electroencephalography (EEG) signals.

Deep learning techniques, popularly used, have multiple layers in their network. Gautam and Sharma (2020) discussed review on deep learning techniques to diagnose neuropsychiatric and neurological disorders. Convolutional neural network (CNN) is a deep neural network technique specifically designed for image processing. It employs convolutional layers to learn hierarchical features from neurological images. It is widely used for brain lesion segmentation in diagnostic imaging scans. AbdulAzeem et al. (2021) achieved high accuracy to classify AD using CNN. There is no doubt that AI, machine learning, and deep learning can outperform human experts in some cases, but deep learning has been very much effective for various diagnostic tasks. While deep learning possesses all these capabilities, it is not used for medical purposes because of its black box nature. Recent deep learning studies are aimed at the decision-making procedure of the algorithm so that they can be used for clinical purposes.

Neurodegenerative diseases and sleep disorders are highly correlated (Famà et al., 2024; Gottesman et al., 2024). Sleep plays an integral part in our lives. Sleep is required not only for the proper functioning of our brain and body but also for a variety of other functions. Sleep uplifts our mood, enhances our cognition, boosts our immunity, and facilitates overall well-being. But on the other hand, improper sleep can lead to a myriad of problems, which can affect both our physical and mental well-being. Sleep apnea is a grave sleep disorder. It is a condition where a person's breathing starts and stops many times while the person is asleep. A person suffering from sleep apnea has difficulty in performing tasks that require cognition. It is mostly found in the individuals who are elderly, around 55–60 years old. Widespread presence of stress and poor sleeping habits are very common in our modern lifestyle. When sleep-deprived people also have a habit of excessive alcohol consumption or heavy smoking or intake high caffeine, they may develop insomnia. Obesity is one of the prominent causes of sleep apnea. Untreated sleep apnea can lead to some serious diseases including cardiovascular diseases like heart attack or stroke and metabolic disorders like type 2 diabetes and neurological disorders like AD. Hence, early detection and treatment of sleep apnea have become a necessity.

In this chapter, we have reviewed various machine learning and deep learning approaches to identify brain abnormalities using neuroimaging analysis and how sleep apnea affects brain adversely and leads to neurodegenerative disorders.

3.2 PREPROCESSING AND FEATURE EXTRACTION IN NEUROIMAGING DATA

Neuroimaging preprocessing, akin to treating sleep apnea in data analysis, is the process of preparing neuroimaging data and making it suitable for machine learning-based analysis by identifying and removing errors or irregularities in the data, reducing noise, etc. This helps in improving the overall quality of the data that is being sent to be analyzed by the machine learning model. Preprocessing steps are as follows.

3.2.1 Motion Correction

Motion correction is a common preprocessing technique applied to neuroimaging data, particularly in the case of fMRI. Extracting functional information from such neurological images involves statistical time-series analysis, which works with the assumption that the location of voxel in the subject's brain is not changing. A voxel is a three-dimensional representation of a discrete brain tissue. Each voxel's intensity contains information about the brain tissue it represents. If the subject is moving, we could risk extracting wrong information. The technique works by selecting a reference image which is used to correct the images with motion blur.

3.2.2 Spatial Smoothing

Spatial smoothing is the process of applying a spatial filter to neuroimaging data with the goal of increasing the signal-to-noise ratio (SNR) and reducing any spatial noise present in the data. Some of the commonly used filters include average smoothing, adaptive smoothing, and Gaussian smoothing. The average smoothing filter operates by taking the average of the neighborhood pixel values. A side effect of this filter is the blurring of edges which are as important as all the other pixels in the image. To reduce the blur, the Gaussian smoothing filter increases the weight of the central pixel and decreases the weight of the distant ones. This helps in bringing the intensity of the new central pixel close to its neighboring pixels.

3.2.3 Data Cleaning

In this stage of the model making, the data is taken from the datasets, and it is cleaned by handling out the missing values in the dataset. The missing values are handled by using mean, median, or mode in place of the missing values or removing rows or columns entirely. Data transformation such as converting numerical types of values into string type values or vice-versa or converting categorical columns into binary values are done to maintain some overall consistency. Once the data is cleaned, it is often split into two categories: one for testing and the other for training the model.

3.2.4 Dimensionality Reduction in Neuroimaging

In the field of neuroscience, the input data usually exhibits much greater dimensionality than the number of training samples. For example, in EEG samples, the dimensionality can range up to 10^5, while the number of samples is typically much smaller, often only a few hundred. In order to make classification tasks feasible, the dimensionality of the input data must be considerably reduced, so that the necessary features can be extracted. There are two main categories of approaches for dimensionality reduction, unsupervised methods, and supervised methods. The unsupervised methods include techniques such as independent component analysis (ICA), principal component analysis (PCA), sparse PCA, or non-negative matrix factorial. Apart from dimensionality reduction, these techniques also help in removing noise from the data and thereby enhancing the SNR. The supervised methods make use of class labels during the learning phase to identify a transformation of input data which has a more compact set of features that are highly relevant to the task at hand (Lemm et al., 2011).

3.2.5 Maintaining Quality of Data

Neuroimaging data is quite complex to understand and often is quite difficult to extract. To maintain a quality dataset, an expert data engineer is required who can remove differences and standardize the data. If the data quality standard is not met, then the poor dataset will result in poor results and may introduce some biases which will directly have an impact on the reliability of the diagnosis. Quality control is the practice of identifying and removing poor quality images caused due to technical or subject-related issues so that they do not compromise the quality of the overall dataset.

3.2.6 Avoid Limited Data

To effectively train a machine learning model, one must supply the model with copious amounts of data. If the supplied data is limited, then it would be very difficult to effectively train the machine learning model and make it predict the results more accurately. Small datasets can also cause the resultant graphs to be less detailed than needed, and it will affect the interpretability of the results.

3.2.6.1 Feature Extraction Techniques

The first step to make a machine learning model that uses neuroimaging analysis to detect sleep apnea is to select some features upon which the model will be trained. The machine learning models, extract feature which are essential for neuroimaging analysis.

Various computer-aided diagnosis tests are available like X-rays, ultrasound, CT scans, MRI, fMRI, and EEG to produce neuroimages, and polysomnography (PSG) graphs recordings provide information about regions of brain activity, sleep trends, snoring severity, formation of tumors, etc. These features provide information about brain activity, sleep trends, snoring severity, etc. MRI scans provide information about the anatomy of a patient's brain which is used to detect sleep disorders. The fMRI scans give insights into any abnormalities of the brain. They also show the various parts of the brain which are actively working. The sleep specialist examines this data to evaluate the severity of sleep apnea. EEG captures brain activity which makes it easier to identify problematic regions in the person's sleep trends. PSG graphs provide a wide variety of data such as respiratory rate, heart rate, SpO2, and bodily movements during sleep.

3.2.7 Magnetic Resonance Imaging

MRI provides us with high-resolution images of our brain including certain patterns and textures which can be analyzed, allowing us to predict possible brain disorders in individuals.

Unlike CAT scan which makes use of ionized radiation to produce neurological images, MRI uses magnetic and radio waves, making it safe for repeated use.

Texture analysis is a technique used to identify spatial patterns of intensity variations in MRI images by assessing local variations in image texture which is later used to differentiate between healthy and pathological tissue.

Voxel based morphometry (VBM) is another neuroimaging analysis technique which generally involves the use of statistical maps that display the regions of considerable difference or correlation across the brain. These regions further serve as biomarkers for disease detection.

3.2.8 Positron Emission Tomography

A PET scan creates 3D images of the body with the help of radioactive tracers, which are typically administered to a patient via intravenous injection. These tracers contain carrier molecules which can combine with our target sugar or proteins present in the body. The tracer emits positrons which interact with surrounding electrons to emit two photons (gamma rays) in opposite directions. By analyzing the distribution and timings of gamma rays, the PET scanner can construct a 3D image representing tracer distribution within the patient's body which can be later assessed to predict possible brain disorders.

Standard uptake value (SUV) is a common PET image analysis technique which is used to calculate the concentration of gamma rays within a region of interest (ROI). It is computed based on the activity concentration within the ROI normalized to the patient's weight and the injection dose of the tracer.

SUV = Injected dose (Bq)/Body weight (g) Activity concentration (Bq/g) (Hirata and Tamaki, 2021).

3.2.9 Electroencephalography

EEG captures brain activity which makes it easier to identify problematic regions in the person's sleep trends. An EEG is used to capture the electric signals from the brain using electrodes and transform them into graphical representation of brain activity w.r.t time and frequency. The waveforms are then assessed by doctors to diagnose neurological disorders. The electrical activity of the brain is represented as waves of varying frequencies, amplitude, and shape. Machine learning models use these EEG graphs for detecting sleep apnea, seizures, and epilepsy.

3.2.10 Functional MRI

The fMRI scans gives insights about any abnormalities of the brain. It also shows the various parts of the brain which are actively working. The sleep specialist examines this data to evaluate the severity of sleep apnea.

3.2.11 Polysomnography

PSG is a multi-parameter sleep study where the patient's sleep is monitored in detail. PSG gives insights into the patients sleep patterns to the sleep specialist by providing a variety of data such as ECG signals, snoring signals, SpO2, and heart rate signals, to name a few.

PSG studies are generally done within a sleep clinic where some electrodes are attached to the patient's scalp and temple. These electrodes relate to a recording machine and collect brainwaves and related data inside it such as EEG signals. Respiratory sensors are attached to the patient's chest which store information related to breathing patterns and oxygen levels (SpO2). Some sensors are even attached to patient's legs and arms to catch any twitching or movement while sleeping. The advantage of PSG study is not restricted to the detection of sleep apnea, but it can also detect restless leg syndrome, cardiac activity, brain activity, oxygen saturation, etc.

PSG graphs provide a wide variety of data such as respiratory rate, heart rate, SpO2, and bodily movements during sleep. The patient may also be asked to record their sleep using their smartphones or fitness tracker bands; they can keep track of their bedtime schedule, snoring sounds, heart rate, blood pressure, stress levels, etc., to identify sleep apnea.

3.3 COMMONLY USED MACHINE LEARNING ALGORITHMS FOR NEUROIMAGING ANALYSIS

Once we have our data cleaned and split into training and testing, the next step is to select algorithms on which the data will be trained. Supervised learning algorithms such as regression and classification algorithms can predict the outputs based on the data supplied to it. If these algorithms are trained on extensive neuroimaging datasets, the model can efficiently predict the factors causing sleep apnea.

Before releasing the model into the real world for real-time analysis, the model should be validated and tested thoroughly. The trained model is tested on the testing dataset which is used to determine its performance, reliability, accuracy, and any biases. The main function of the validation of the model is to make sure whether the model created is generating outputs according to intended functionality. Once the model is validated, the machine learning model can be applied to medical fields and perform real-time analysis. Health professionals and sleep analysts can use these models to speed up and automate some parts of the diagnosis process. The models are often linked with applications which users can install in their smart devices and take care of their health in the comfort of their home.

In this section, we will discuss some of the common machine learning algorithms that are used for analysis of neurological images.

3.3.1 Linear Discriminant Analysis

Linear discriminant analysis is a common technique used for dimensionality reduction problems. Dimensionality reduction serves as a pre-processing step in machine learning to reduce the number of features. It can easily handle the cases for unequal frequencies in a class, and performance is determined based on randomly generated data. LDA provides better classification than PCA. PCA focuses more on feature classification, and LDA does data classification (Balakrishna and Ganapathi Raju, 1998; Tharwat et al., 2017).

3.3.2 Support Vector Machines

SVM is a supervised learning algorithm which can perform both classification and regression tasks, but it is most used for classification. One significant advantage of SVM is that they are exceptionally useful while dealing with high-dimensionality datasets (such as neuroimages), where the number of features far exceeds the number of data samples. In the past, SVMs had been widely used for generating diagnostic scores based on brain images. SVMs use information derived from multivariate imaging to make informed decisions (Gaonkar et al., 2015).

According to a recent paper, "[t]he combination of SVM with other methods allows to improve feature selection and to avoid overfitting of data, and this will facilitate the generalization of the model (i.e., achieving high accuracy when applied to different datasets)" (Grueso and Viejo-Sobera, 2021).

3.3.3 Random Forest

RF is another supervised learning algorithm which can solve both classification and regression problems. It is an ensemble learning method that combines the result of multiple DTs to generate an output.

Its ability to handle highly non-linearly correlated data, resilience to noise, and flexible nature makes it an ideal candidate for the analysis of high-dimensional data (such as neurological images) (Sarica et al., 2017).

3.3.4 Convolutional Neural Networks

CNN is a type of neural network architecture that is commonly employed in computer vision which enables machines to understand and process structured grid-like visual datasets, such as images or videos. In recent years, CNN has attracted a lot of attention due to its ability of classifying and analyzing images. CNN consists of many layers such as input layer, convolutional layer, max pooling layer, fully connected layer, and an output layer.

Neurological images, such as MRIs or fMRIs, are generally represented as 2D or 3D arrays of voxel intensities. Each voxel is a small volume element in the brain, and its intensity represents the activity level in that region of the brain. The convolutional layer consists of filters (also known as kernels) which capture features or spatial patterns, such as textures or anatomical landmarks from neurological data to produce a feature map. The pooling layers help in reducing the dimensionality of the extracted feature map while retaining the important information. This step helps us to reduce the complexity of computation and the risk of overfitting. After the above step, activation functions, such as ReLU (Rectified Linear Unit), are applied which enable CNN to learn complex relationships from the extracted features. Using the convolutional layers, we can use the extracted features to make predictions about specific outcomes, such as disease diagnosis.

In the past, CNN has been successful in predicting mild cognitive impairment (MCI)-to-AD with an accuracy of 79.9% and an area under the receiver operating characteristic curve (AUC) of 86.1%. MCI is the early stage of AD as suggested by Lin et al. (2018) in the past.

3.3.5 Decision Tree

In this algorithm, a graph is used to represent results and choices in the form of tree. The nodes in the tree represent a choice, and the edges represent rules based on which the decisions are made. This algorithm can provide a better and clearly understandable structure in comparison with other machine learning algorithms. The decisions taken based on this algorithm can be used by physicians in making predictions (Balakrishna and Ganapathi Raju, 1998; Sahithi and Sripad, 2018).

3.3.6 K-means Clustering

It is the simplest unsupervised machine learning algorithm which can solve clustering problems. It involves a simple and easy way to classify certain data through a certain number of clusters. Being a simple and efficient approach, this algorithm can be used for segmentation tasks. This will provide a clear visualization of the images where each region in an image will be referring to cluster which will eventually help getting a clear visualization (Mahesh, 2020).

3.3.7 Deep Learning Techniques

Long short-term memory (LSTM) was primarily proposed by Hochreiter and Schmid Huber in 1977. LSTM is a sort of recurrent neural networks (RNNs) can efficiently solve the vanishing gradient problem. Abed and Ibrikci (2023) used gradient decent to determine when to save and when to retrieve the information.

Another technique came with similar concepts like the LSTM and Gated Recurrent Unit (GRU) but with a small number of parameters which help in training faster than LSTM. Studies showed that LSTM and GRU are comparable in some of the cases. The similarities are that they both use a similar internal structure.

Residual network (ResNet) is a type of already trained deep learning models which are used for feature extraction, prediction, or specially tuned for a specific case. They input the learnings from previous findings and use them for new problems to solve them quickly. The ResNet50 deep learning model was used for detecting sleep apnea.

Later, a combination of ResNet50 architecture along with additional two layers was used, which is called Hybrid Model. One LSTM layer precedes the output layer, and one GRU layer follows the input layer.

3.3.7.1 Deep Learning Implementation

The use of deep learning for applying facial depth diagnosis for obstructive sleep apnea (OSA) was developed. Human facial scans provide more additional information than 2D color pictures, thus providing additional information about the facial features of an individual. The Human Research Ethics Committee approved this study. The models, such as VGG19, VGG face, PAMs-Alex, and Pose aware models (PAMs), were used for training facial identification features. Out of these models, VGG outperformed Res. models with an accuracy of 69%.

Deep learning models are used to develop an engaging method for detecting sleep disorders. Hepsiba et al. (2021) used the electrooculography (EOG)signal that estimates the amount of activity in the eye which gives us information about the sleep disturbances which are used in disorder detection. It basically uses six scores (sleep status [from stage 1 to stage 4], wake status and rapid eye movement (REM) for the analysis of the database. Determination of GRU in this computerized model of sleep disorder prediction is done using advanced deep learning techniques.

Insomnia diagnosis can also be done with the help of deep learning, and it was carried out at the Interdisciplinary Sleep Medicine Center at Charité-Universitätsmedizin Berlin. In this study, information of 41 normal individuals and 42 insomnia patients was used for

PSG evaluation, and PSG deep learning was applied on separate insomnia patients. Ravyts et al. (2021) used DNN and DNN-HMM classifiers to show that deep learning would be more beneficial in the detection of insomnia.

In the future, smaller sensors-based devices will be used which could help us in even detecting the smallest possible problems and home-based monitoring systems which would be cost-effective and will entirely change sleep-related research (Mohan et al., 2023).

3.3.8 Explainable Deep Learning

The increase in the growth of AI, the computing power, and the large amounts of data has seen extensive development in the past few years, and various AI software tools are widely used in our daily lives. Even the AI-powered systems are of great advantage to us, but still their black box nature makes them opaque, and their decision-making process is not known and therefore questions arise about the explanation for the decisions taken by them. This led to the introduction of explainable artificial intelligence (XAI), which promotes AI algorithms which enable machines to learn, analyze, and make decisions, and XAI provides an explanation for the decision-making process (Minh et al., 2022).

Detection of sleep disorders is always considered to be a complex task, and earlier PSG was used but with the help of deep learning algorithms. It is possible to automate the task of sleep stage classification which can be used as an input to the sleep disorder detection process. The first approach uses various features like time frequency representation (TFR), the Fourier-Bessel decomposition method (FBDM), and a CNN classifier to categorize EEG signals. It is possible to extract features and identify the sleep stage using machine learning approaches.

Older techniques for determining sleep disorder used 2D CNN models for their research purposes which left space for the explanation of medical time series. Hepsiba et al. (2021) have suggested a model that helps in automatically locating distinct retinopathy lesions using original diabetic retinopathy (DR) funds images. Lee and Kim (2018) proposed a recurrent-convolution neural network of EEG-based intention recognition. Unnecessary channels were removed, and essential was selected using Grad-CAM.

Explainable machine learning is focused on determining those regions of the signals which are most relevant to get a right prediction from the model. One such visualization technique is Grad-CAM. This technique is mostly used as an image classifier and is used for classification tasks. The prediction of the model is presented in such a way that it can be understood by the clinician. The aim of developing an explainable deep learning model in detecting is to detect OSA and compute the AHI value to compute the severity so that it can be used in pre monitoring (PMs) in healthcare centers. So, for a clinical setup, a good interpretability of the model is required.

3.3.9 CNN+RNN

Various simplified diagnostic tests have been developed for the detection of OSA, and among all of them AF and SpO2 have been used in all the alternatives developed for PSG. Both AF and SpO2 are involved in apnea and hypopnea definitions. A more simplified approach for OSA detection is focused on machine learning, but it is dependent on feature extraction. However, these extra efforts can be eliminated with the use of emergent deep learning which requires no feature extraction as it automatically does it by finding co-relative data patterns in the data (Jiménez-García et al., 2024).

Deep learning-based approaches for the detection of OSA require the use of CNNs and RNNs and have shown their effectiveness in automatically extracting complex patterns from the dataset. RNNs help to model the temporal structure of the patterns extracted. To get the benefit of both, CNN+RNN are used together.

Integrating CNN and RNN helps in the formulation of AF and SpO2 data for detecting OSA. Finally, the Grad-CAM will help in understanding the OSA disorder detection working of the deep learning model.

Explainability methods for deep neural networks (DNN): approaches used to explain the use of DNN in medical imaging.

Singh et al. (2020) used standard attribution-based methods and the other uses domain-specific architecture. Most studies conducted for explainable deep learning in clinical image diagnostic use feature-based techniques. The model has a ready-to-use nature. Open-source implementation is available to make a solution easy and convenient to use. By doing this, a person using deep learning for medical imaging can focus on building a model which is best suited for a given task, and this will eventually help in getting a better understanding of the model. In the classification problem, the main reason for using standard attribution methods is to discover the usefulness of an input attribute to the recipient neuron which is typically the final neuron that predicts the accurate class.

3.4 SLEEP DISORDERS AND NEURODEGENERATIVE DISEASES

Obstructive sleep apnea is the most common form of sleep apnea. A majority of patients suffering from sleep apnea have OSA. It occurs when a person is lying in a supine position. The lower jaw falls back due to gravity, causing a restriction in the airflow to reach the lungs. Obstructive sleep apnea is commonly observed in individuals who are overweight. Loud snoring, episodes of choking during sleep, and sore throat are some symptoms of OSA.

Another very common form of sleep apnea is **Central** sleep apnea in which a person's airway path is not restricted. It occurs when the brain fails to transmit the signals to control the muscles which are responsible for breathing. Central sleep apnea is commonly observed in people with neurological disorders like Parkinson's disease and amyotrophic lateral sclerosis (ALS). People suffering from central sleep apnea may experience uneven heartbeat, palpitations, and high blood pressure. These individuals have a higher risk of getting a stroke, diabetes, or obesity.

Neuroimaging analysis presents us with distinctive results in predicting various disorders, such as Alzheimer's, dementia, and Parkinson's.

3.4.1 Neurodegenerative Disorders in Neuroimaging

3.4.1.1 Parkinson's Disease

Neuroimaging can be used in the diagnosis of Parkinson's disease as it helps in differentiating it from other conditions which show its symptoms. Dopamine tracer imaging techniques like single-photon emission tomography (SPECT) and PET are used to assess the dopamine synthesis in the brain. Neuroimaging techniques alone might not be able to provide appropriate diagnosis report, but with the use of MRI and dopamine tracer imaging, we can surely get better results (Risacher and Saykin, 2013; Hepsiba et al., 2021).

3.4.1.2 Alzheimer's Disease

It is a common problem in the elderly and is associated with loss of neurons from certain parts of the brain. Neuroimaging can provide insights about the brain structure and its function. It helps in differentiating Alzheimer's from other age-related cognitive decline disorders. With the help of structural MRI, doctors can get knowledge of any shrinkage of the hippocampus which is crucial for the memory. PET scans are mainly used for measuring brain glucose metabolism as in the cases of AD, there is reduced glucose metabolism in several parts of the brain (Shimizu et al., 2018).

3.4.1.3 Huntington's Disease

This is a neurodegenerative disorder which is inherited, and it affects the nerve cells inside the brain. It becomes worse over time, being a progressive disease. It is believed that this disease starts between the ages of 30–50 years. The common symptoms are having problem in body movements like incoordination between body parts and difficulty in walking. Neuroimaging is proved to be very useful in detecting this disorder in early stages with the help of imaging techniques like PET and SPECT. Physicians can get important information about the activity of the brain and analyze changes in an individual having the disease. With this extensive capability of neuroimaging, it is possible to build better treatment methodologies for the treatment of the disease (Risacher and Saykin, 2013).

3.4.1.4 Cerebral Blood Flow and Glucose Metabolism

In neuroimaging, the fluorodeoxyglucose positron emission tomography (FDG-PET) is used to get an idea about both regional and cerebral blood flow (RCBF) and metabolism. FDG-PET is a neuroimaging technique to measure the activity of brain and determines glucose metabolism. Inside the brain, the active regions have a larger demand of glucose for more energy, and FDG-PET detects the accumulation of this radiotracer in the prominent areas. With the help of FDG, we can get information about the level of activity in different regions of the brain. By diagnosing patterns of reduced metabolism in certain parts of the brain, this technique helps in the early detection and diagnosis of Huntington's disease (Risacher and Saykin, 2013; Shimizu et al., 2018).

Disturbed sleeping patterns are quite common these days due to the increasing level of stress, late night lifestyle routines, and eating habits. Moreover, when these patterns persist for longer durations in an individual, they may result in sleep apnea and insomnia problems. Sleep disorders in long run may lead to some serious diseases including cardiovascular diseases like heart attack or stroke, metabolic disorders like type 2 diabetes, and neurological disorders like AD. Therefore, it is desired to detect sleep apnea at an early stage. At present, sleep apnea cases are very common, so there is a need to devise a cost-effective and convenient mechanism for the early detection and treatment of sleep apnea.

With evolving time, the most popular technique to gather information about patients was polysomnogram test, a multi-parameter sleep study which is considered a gold standard for detecting sleep disorders as it has good accuracy and highly effective but is hard to conduct as overnight detailed observation is required.

Many studies were conducted for the detection of sleep apnea events diversity. Sleep apnea can be detected by various techniques like Epworth Sleepiness Scale, which measures the patient's daytime sleepiness questionnaire and Home Sleep Apnea test (HSA), a portable device that helps the person to monitor sleep parameters such as total sleep time, number of awakenings, any sort of respiratory disturbance, etc. The results obtained by using HST include Apnea-Hypopnea Index (AHI), Respiratory Disturbance Index (RDI), and Oxygen Desaturation Index (ODI).

Firstly, raw data is transformed into features that are suitable for machine learning models; afterwards, machine learning/deep learning techniques are applied.

OSA is a very common condition, and it is estimated to affect 425 million adults aged between 30 and 69 years. In recent years, home testing for OSA detection has increased to a larger extent, and it is an alternative to PSG. The statistical combination of the results of multiple studies revealed that the misdiagnosis rate of OSA at home is 39%. These factors lead to the development of portable diagnostic techniques like the one based on single channel oximetry analysis (Levy et al., 2023).

Awareness for sleep apnea has increased but still a large number of people are undiagnosed because of the poor availability of cost-effective/less time-consuming alternatives instead of the PSG test. Meanwhile, some studies have developed OSA screening tests at two levels: OSA and non-OSA. Some took into account multiclass classification tasks which considered all cases from mild to severe, but they failed in estimating the AHI.

Levy et al. (2023) developed deep learning approach known as OxiNet, in various distributions. OxiNet has set a benchmark in identifying OSA. OxiNet missed only 0.2% test cases among all the test sets for moderate-to-severe OSA.

Overall, 12,923 PSG records from six different databases were used to construct and assess OxiNet to perform regression tasks of AHI approximation from the oximetry signal. According to the American Academy of Sleep Medicine (AASM) 2012 rules and ICSD-3 guidelines, the AHI is described as the average number of all apnea and hypopnea events that occur when a person is sleeping.

The researchers of Pennsylvania State University designed a smartwatch-based system called Apnea Detector to identify the presence of sleep apnea in individuals. They managed to get a remarkable f1-score of 96.49% (Chen et al., 2021).

3.5 DISCUSSION

This chapter gives an overview about machine learning and deep learning algorithms applied on neuroimages and predicts various neurodegenerative disorders. The initial phase of the chapter dives into neuroimaging, emphasizing the foundational aspects and utilized methodologies. It signifies the importance of using machine learning in neuroimaging analysis, showcasing usefulness in data interpretation and decision-making procedures. Prominence is given to preprocessing of neuroimaging data to maintain optimal quality and reliability. The assessment also reveals the indispensable role of dimensionality reduction techniques in selecting the most prominent features for the analysis. It sheds light on feature extraction techniques used for neuroimaging.

A complete assessment is conducted to observe frequently used algorithms for analyzing neuroimages. Transitioning to sleep medicine, providing an extensive overview of sleep apnea, Li et al. (2024) used SVM to classify patients with mild cognitive impairment.

Traditional techniques such as the Epworth Sleepiness Scale and PSG among others are discussed. Modern diagnosis systems including health applications and non-intrusive wearable devices are also reviewed.

In the view of machine learning model development, this phase of the chapter delves into the investigation of highlighting the pivotal role of algorithm selection and model validation. Moreover, it investigates the challenges inalienable in utilizing neuroimaging information for symptomatic purposes, emphasizing the requirements for strong strategies and interpretability.

The final phase of the chapter draws attention to different deep learning techniques used for sleep apnea detection, utilizing CNNs and RNNs for boosting the performance of the system. The prominence of explainable deep learning models is underlined, especially in the field of medicine using image analysis techniques where interpretability is paramount. It explores different techniques such as Grad-CAM visualizations to explain how deep neural networks reach at their decisions, therefore promoting model interpretability and transparency.

The chapter highlights the interconnection between neuroimaging, sleep medicine, and machine learning. It gives an overview of innovative approaches, confronts current hurdles, and paves a path for future advancements in these areas.

3.6 CONCLUSION

In conclusion, the chapter highlights the integration of machine learning algorithms into neuroimaging analysis and diagnosis. It signifies the importance of machine learning in data interpretation and decision-making. The chapter explores the importance of dimensionality reduction techniques and various machine learning algorithms frequently used for neuroimage analysis. Moreover, it provides a thorough rundown of sleep apnea, inclusive of traditional and modern techniques of diagnosis. The chapter also discusses the challenges faced in diagnosis using neuroimaging and lays emphasis on the need for more robust approaches. In the end, it focuses on the use of machine learning and deep learning techniques for sleep apnea detection, underlining the importance of explainable models in neuroimaging analysis. In summary, the chapter showcases the relationship between neuroimaging, sleep apnea, and machine learning, unveiling new strategies that set the stage for further advancement in these domains.

REFERENCES

AbdulAzeem, Y., Bahgat, W. M., & Badawy, M. (2021). A CNN based framework for classification of Alzheimer's disease. *Neural Computing and Applications*, *33*(16), 10415–10428.

Abed, M., & Ibrikci, T. (2023). Sleep apnea events detection using deep learning techniques. *Journal of Sleep Disorders: Treatment and Care*, *12*(1), 2.

Balakrishna, S., & Ganapathi Raju, A. (1998). Linear discriminant analysis—A brief tutorial. *Institute for Signal and Information Processing*, *18*(1998), 1–8.

Chatterjee, I., Baumgartner, L., & Cho, M. (2023). Detection of brain regions responsible for chronic pain in osteoarthritis: An fMRI-based neuroimaging study using deep learning. *Frontiers in Neurology*, *14*, 1195923.

Chatterjee, I., Kumar, V., Sharma, S., Dhingra, D., Rana, B., Agarwal, M., & Kumar, N. (2019). Identification of brain regions associated with working memory deficit in schizophrenia. *F1000Research*, *8*.

Chen, X., Xiao, Y., Tang, Y., Fernandez-Mendoza, J., & Cao, G. (2021). Apneadetector: Detecting sleep apnea with smartwatches. *Proceedings of the ACM on Interactive, Mobile, Wearable and Ubiquitous Technologies*, *5*(2), 1–22.

Famà, F., Arnulfo, G., Arnaldi, D., Kim, H. J., & Jung, K. Y. (2024). EEG-based machine learning models for the prediction of phenoconversion time and subtype in isolated rapid eye movement sleep behavior disorder. *Sleep*, *20*, 1–10.

Gaonkar, B., Shinohara, R. T., Davatzikos, C., & Alzheimers Disease Neuroimaging Initiative. (2015). Interpreting support vector machine models for multivariate group wise analysis in neuroimaging. *Medical Image Analysis*, *24*(1), 190–204.

Gautam, R., & Sharma, M. (2020). Prevalence and diagnosis of neurological disorders using different deep learning techniques: A meta-analysis. *Journal of Medical Systems*, *44*(2), 49.

Gottesman, R. F., Lutsey, P. L., Benveniste, H., Brown, D. L., Full, K. M., Lee, J. M., J.-M., Osorio, R. P., Matthew, P., Redeker, N. S., Redline, S., American Heart Association Stroke Council, Council on Cardiovascular and Stroke Nursing, & Council on Hypertension. (2024). Impact of sleep disorders and disturbed sleep on brain health: A scientific statement from the American Heart Association. *Stroke*, *55*(3), e61–e76.

Grueso, S., & Viejo-Sobera, R. (2021). Machine learning methods for predicting progression from mild cognitive neuromimpairment to Alzheimer's disease dementia: A systematic review. *Alzheimer's Research & Therapy*, *13*, 1–29.

Hepsiba, D., Anand, L. V., & Princy, R. J. P. (2021, March). Deep learning for sleep disorders: a review. *2021 seventh international conference on bio signals, images, and instrumentation (ICBSII)* (pp. 1–5). IEEE.

Hirata, K., & Tamaki, N. (2021). Quantitative FDG PET assessment for oncology therapy. *Cancers*, *13*(4), 869.

Jiménez-García, J., García, M., Gutiérrez-Tobal, G. C., Kheirandish-Gozal, L., Vaquerizo-Villar, F., Álvarez, D., Del Campo, F., Gozal, D., & Hornero, R. (2024). An explainable deep-learning architecture for pediatric sleep apnea identification from overnight airflow and oximetry signals. *Biomedical Signal Processing and Control*, *87*, 105490.

Kumar, R., Maheshwari, S., Sharma, A., Linda, S., Kumar, S., & Chatterjee, I. (2024). Ensemble learning-based early detection of influenza disease. *Multimedia Tools and Applications*, *83*(2), 5723–5743.

Lee, W., & Kim, Y. (2018, July). Interactive sleep stage labeling tool for diagnosing sleep disorder using deep learning. In *2018 40th annual international conference of the IEEE engineering in medicine and biology society (EMBC)* (pp. 183–186). IEEE.

Lemm, S., Blankertz, B., Dickhaus, T., & Müller, K. R. (2011). Introduction to machine learning for brain imaging. *NeuroImage*, *56*(2), 387–399.

Levy, J., Álvarez, D., Del Campo, F., & Behar, J. A. (2023). Deep learning for obstructive sleep apnea diagnosis based on single channel oximetry. *Nature Communications*, *14*(1), 4881.

Li, Y., Shao, Y., Wang, J., Liu, Y., Yang, Y., Wang, Z., & Xi, Q. (2024). Machine learning based on functional and structural connectivity in mild cognitive impairment. *Magnetic Resonance Imaging*, *109*, 10–17.

Lin, W., Tong, T., Gao, Q., & Qu, X. (2018). Convolutional neural networks-based MRI image analysis for the Alzheimer's disease prediction from mild cognitive impairment. *Frontiers in Neuroscience*, *12*, 412254.

Mahesh, B. (2020). Machine learning algorithms—A review. *International Journal of Science and Research*, *9*(1), 381–386.

Mateos-Pérez, J. M., Dadar, M., Lacalle-Aurioles, M., Iturria-Medina, Y., Zeighami, Y., & Evans, A. C. (2018). Structural neuroimaging as clinical predictor: A review of machine learning applications. *NeuroImage: Clinical*, *20*, 506–522.

Minh, D., Wang, H. X., Li, Y. F., & Nguyen, T. N. (2022). Explainable artificial intelligence: A comprehensive review. *Artificial Intelligence Review*, 1–66.

Mohan, M., Veena, G. N., Pavitha, U. S., & Vinod, H. C. (2023). Analysis of ECG data to detect sleep apnea using deep learning. *Journal of Survey in Fisheries Sciences*, *10*(4S), 371–376.

Nour, M., Senturk, U., & Polat, K. (2023). Diagnosis and classification of Parkinson's disease using ensemble learning and 1D-PDCovNN. *Computers in Biology and Medicine*, *161*, 107031.

O'Donnell, L. J., & Schultz, T. (2015). Statistical and machine learning methods for neuroimaging: Examples, challenges, and extensions to diffusion imaging data. *Visualization and processing of higher order descriptors for multi-valued data* (pp. 299–319). Springer International Publishing.

Quaak, M., van de Mortel, L., Thomas, R. M., & van Wingen, G. (2021). Deep learning applications for the classification of psychiatric disorders using neuroimaging data: Systematic review and meta-analysis. *NeuroImage: Clinical*, *30*, 102584.

Ravyts, S. G., Dzierzewski, J. M., Perez, E., Donovan, E. K., & Dautovich, N. D. (2021). Sleep health as measured by RU SATED: A psychometric evaluation. *Behavioral Sleep Medicine*, *19*(1), 48–56.

Risacher, S. L., & Saykin, A. J. (2013, September). Neuroimaging biomarkers of neurodegenerative diseases and dementia. *Seminars in Neurology*, *33*(04), 386–416.

Sahithi, P., & Sripad, D. V. (2018). Answering to a mysterious question—"Is time travel possible?" *International Journal of Science and Research*, *7*, 426.

Sarica, A., Cerasa, A., & Quattrone, A. (2017). Random forest algorithm for the classification of neuroimaging data in Alzheimer's disease: A systematic review. *Frontiers in Aging Neuroscience*, *9*, 329.

Shahid, S., & Iftikhar, S. (2022). Machine learning models to predict neuropsychiatric disorders in various brain tumors. *Current Medical Research and Opinion*, *38*(5), 687–696.

Shimizu, S., Hirose, D., Hatanaka, H., Takeshita, N., Kaneko, Y., Ogawa, Y., Sakurai, H., & Hanyu, H. (2018). Role of neuroimaging as a biomarker for neurodegenerative diseases. *Frontiers in Neurology*, *9*, 342547.

Singh, A., Sengupta, S., & Lakshminarayanan, V. (2020). Explainable deep learning models in medical image analysis. *Journal of Imaging*, *6*(6), 52.

Tharwat, A., Gaber, T., Ibrahim, A., & Hassanien, A. E. (2017). Linear discriminant analysis: A detailed tutorial. *AI Communications*, *30*(2), 169–190.

Tuncer, T., Dogan, S., Baygin, M., Barua, P. D., Palmer, E. E., March, S., Ciaccio, E., Tan, R., & Acharya, U. R. (2024). FLP: Factor lattice pattern-based automated detection of Parkinson's disease and specific language impairment using recorded speech. *Computers in Biology and Medicine*, *173*, 108280.

Xiao, P., Li, Q., Gui, H., Xu, B., Zhao, X., Wang, H., Tao, L, Chen, H., Wang, H., Tao, L., Chen, H., Wang, H., Lv, F., Luo, T., Cheng, O., Luo, J., Man, Y., Xiao, Z., & Fang, W. (2024). Combined brain topological metrics with machine learning to distinguish essential tremor and tremor-dominant Parkinson's disease. *Neurological Sciences*, *45*, 4323–4334.

Zhang, L., Wang, M., Liu, M., & Zhang, D. (2020). A survey on deep learning for neuroimaging-based brain disorder analysis. *Frontiers in Neuroscience*, *14*, 560709.

4 Precision Medicine in Neurodegenerative Disorders

Sourav Ghosh, Sejuti Ray Chaudhary, Monosiz Rahaman, Rajendra Prasad Chatterjee, and Shilpa Chatterjee

4.1 INTRODUCTION

Precision medicine (PM) has a promising beneficial weapon for both doctors and scientists in recent era. However, the precise opportunity and technical theoretic background of PM are complicated and defy rigid borders. Despite the historic launch of the US Precision Medicine Initiative (PMI) in 2015 (Collins and Varmus, 2015), the process of PM needs to be implemented at the discrete stage and transferred across one ailment to another remains a point of contention. PM is described as treatment and prevention techniques that take precise erraticism into account (Collins and Varmus, 2015). Despite this apparent clarity, biomedical groups are dealing with the employment of transformative programmes in practical contexts, as well as whether previously recognized illness entities require redefinition.

Pharmacological treatment for intricate brain illnesses is quite restricted; in the instance of brain proteinopathies, basic neurodegenerative diseases (NDDs), authorized therapies have primarily been medications with limitation of time effectiveness and considerable individual inconsistency in answer. Furthermore, no illness prevention techniques exist. These difficulties emphasize the requirement for a proof-driven rethink of present medical philosophy and tactics to produce successful biomarker-specific beleaguered and modifying medications, as well as effective screening, early detection, diagnostic, and treatment algorithms. Individual differences in the underlying biology emerge from these complex systems dynamics, even when syndromic and symptomatic phenotypes are comparable. A paradigm of PM is critical for addressing unmet requirements in neurological disorders, which frequently have fewer effective treatments and are an increasing cargo on medical systems and individuals (Green et al, 2020; Hess et al, 2021; Kesselheim et al, 2015; Yu and Saykin, 2021).

The aim of PM is to transport beleaguered and programmed interventions custom-made to a patient's molecular transporters of disease. Neural diseases are auspiciously matched representations for PM because of the quickly increasing inherited phenotypic classification and information base, as well as the progress of biomarkers and adjusting treatments.

4.2 NEUROGENERATIVE DISORDERS: OVERVIEW

NDDs are categorized by the loss of neurons and discerning dysfunction associated with medically different proteins that withdraw in the human brain with peripheral organs (Figure 4.1). Hallmark for neurodegenerative disease is mainly cytoskeletal anomalies,

DOI: 10.1201/9781032711102-4

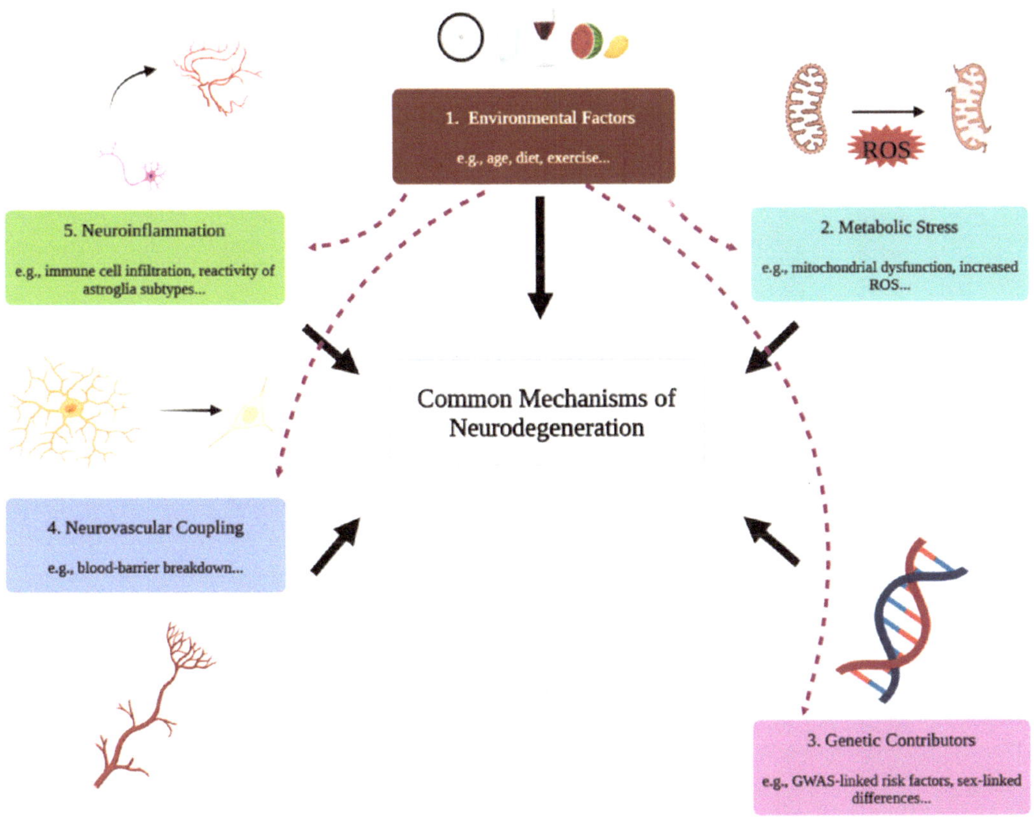

FIGURE 4.1 Etiological framework of neurodegenerative disease.

synaptic and neuronal network defects, DNA & RNA defects, inflammation, neuronic death, and pathological protein accumulation (Lamptey et al, 2022).

4.2.1 Alzheimer's Disease

Recent Alzheimer's disease (AD) pathophysiology research has revealed that amyloid-beta (A) and tau protein buildup is critical to AD development (Bayer and Wirths, 2010; Pickett et al, 2019). The development of protein plaques inside the brain, with neurofibrillary tangles (NFTs) formed of tau phosphorylation, is recognized as a typical hallmark of Alzheimer's (Butterfield et al, 2002; de Paula et al, 2009; Mondragón-Rodríguez et al, 2012). Plaque formation alters hippocampal regions, resulting in poor temporary memory alliance into longstanding traces (Spires-Jones and Hyman, 2014). In AD, there is widespread neuronic loss, incorrect synaptic associations, and impairment to the critical neurotransmitter networks required for brain functioning, as well as memory. Thus, selective memory impairment is the most prevalent scientific sign in early-stage AD.

4.2.2 Parkinson's Disease (PD)

Parkinson's disease (PD) is a deteriorating neurological ailment characterized by muscular rigidity, tremors, a shaky gait, and difficulty with balance and coordination. PD is caused by both nongenetic factors and hereditary. Age is regarded as the key danger factor for PD (Kouli et al, 2018; Spires-Jones et al, 2017). Furthermore, various additional variables, such as extreme coffee use, experience with environmental contaminants, and smoking are recognized to influence the likelihood of developing PD (Priyadarshi et al, 2001), while the precise process is unknown (Emamzadeh and Surguchov, 2018; Gorell et al, 2004; Kuopio et al, 1999). The pathogenesis of PD is characterized by frontal brain atrophy and ventricular hypertrophy. The complete absence of dopaminergic (DA) neuromelanin-containing neurones results in pigment destruction in the substantia nigra pars compacta (SNpc) and position coeruleus (Bartels and Leenders, 2009).

4.2.3 Amyotrophic Lateral Sclerosis (ALS)

ALS, more normally referred to as Lou Gehrig's disease or motor nerve cell disease, is an advanced disease of the spinal cord and neuron, subsequent in muscle dimness and immobility (Morris, 2015; Wang et al, 2020). In this disease, motor nerve cells progressively decline before their dismissal (Hardiman et al, 2017). When motor neurons become damaged, the transmission of signals to the brain is hindered. Despite more than 30 different genes being associated with ALS, changes in four key genes (SOD1, C9orf72, FUS, and TARDBP) are responsible for more than 70% of cases of lateral sclerosis (Hardiman et al, 2017). These four genes encode proteins involved in critical motor functions like mitochondrial function, DNA repair, maintaining homeostasis, and the role of glial cells.

4.2.4 Huntington's Disease

Huntington's disease is another neurodegenerative disorder caused by heredity. It mainly occurs due to the breakdown of neurons in the brain and gene mutation on chromosome 4 of huntingtin (HTT). It is generally associated with behavioural changes and cognitive impairments.

4.3 PRECISION MEDICINE: EMERGING THERAPY

PM is a new translational science paradigm that strives to improve the effectiveness of disease prevention by focusing on the revelational built, multipart of the health-disease equilibrium and range. It employs time-sensitive diagnosis and dealing procedures that are personalized to the patient's genetic–clinical–biological traits, psychosocial atmosphere, and danger factors. A comprehensive healthcare strategy like this is doable. Only a thorough grasp of the disease's biological routes and the documentation of at-risk groups will suffice. Subsequent to this perilous stage of research, the creation of treatments that are pathway-based and stage-dependent targets crucial causative variables and upriver. It is possible to achieve molecular and cellular changes. Eventually, PM-focused methods are hoped to likely result in the efficient incorporation of other than pharmacological (i.e., lifestyle allied) therapies, pharmacological therapy tailored to the person for primary prevention as well as secondary prevention and therapy. PM follows four principles: stratification of each participant or patient based on the risk factors; proposed treatment based on the social and physical; biological characteristics, large screening and therapeutics; and optimizing the advantages of patients via utilization of the self-assessments (Hood and Friend, 2011).

4.4 GENETICS AND PRECISION MEDICINE

Various instruments are utilized to gather these ground-breaking data sources, with genetic data being especially valuable (Figure 4.2). Information from the genome, transcriptome, proteome, and epigenome has garnered significant value and focus in the treatment of neurodegenerative disorders. The easy path to examine the genetical basis of neurological disease. Mutations of PSEN1 and PSEN2, APP genes caused AD and finding novel genes in AD, candidate gene-based and classic linkage-based association lessons have been displaced by exome sequencing research and genome-wide association studies (GWAS) (Harold et al, 2009).

Rapid advancements in sequencing technology have significantly contributed to genetic discoveries in neurological diseases. The three primary categories of genetic information of genome, transcriptome, and epigenome can be effectively explored through various

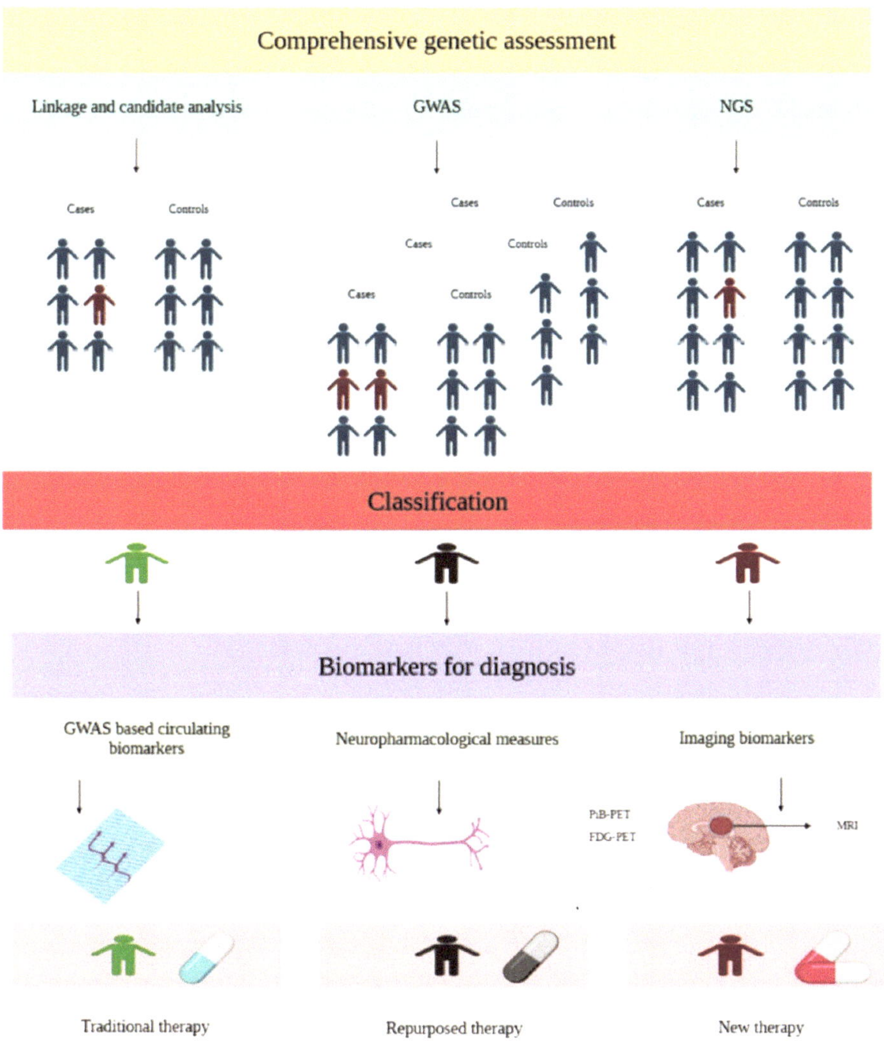

FIGURE 4.2 Role of precision medicine and genetics in disease diagnosis.

assays such as next-generation sequencing (NGS) and GWAS. While the vast number of genes associated with neurological diseases makes individual sequencing for analytical purposes impractical, NGS allows for the concurrent screening of an extensive set of genes linked to these conditions. Even with a small number of patients, NGS approaches not only identify novel risk factors but also provide valuable insights into disease mechanisms. Understanding how nerve cells change and preserve their molecular signatures in disorder progression is a crucial objective in neurology research. ABCA7 (Vardarajan et al, 2015), followed by BIN1 (Tan et al, 2014a), CR1 (Ma et al, 2014), PLD3 (Cacace et al, 2015), and PICALM (Jiang et al, 2014) genes have been pinpointed as carrying an excessive load of harmful coding mutations in AD. In the context of epileptic encephalopathies, trio sequencing has revealed genes such as DNM1, HCN1, ALG, GABRB3, SCN8A, GABRA1, KCNT1, GNAO1, SCN2A, SLC35A2, and GRIN2A associated with epilepsy, many of whose encoded proteins are involved in synaptic transmission (EpiPM Consortium, 2015). NGS technology is revolutionizing the exploration of the molecular foundation of neural function. RNA-sequencing (RNA-seq) delivers detailed data on transcript construction, enabling the inspection of non-coding RNAs, protein-coding genes, and splicing. In AD, RNA-seq in brain tissue facilitates the study of differential transcription, uncovering networks that may contribute to disease mechanisms. Noncoding RNAs observed in animal models and blood samples further contribute to understanding AD progression (Tan et al, 2014b). NGS can be described as the comprehensive molecular signatures of cells via transcriptome analyses including the events that preserve such signatures over epigenetic studies. Epigenetic deviations, involving alterations to DNA molecules and DNA's packaging proteins themselves via not altering the fundamental sequence, are believed to be influenced by the environment, potentially leading to pathogenic alterations in gene expression. Studies on differentially methylated regions afford persuasive evidence for momentous associations with neurological diseases, suggesting a potentially causative relationship. Overall, these findings encourage scientists to focus on rare alternatives in neurological diseases, especially as expansions in NGS facilitate the inclusive exploration of genomes. So, the combined use of GWAS and NGS holds great promise for personalized treatment approaches in neurological disorders, chiefly as therapy targeting the original molecular origins rather than just medical symptoms turn out to be available (Mazzio and Soliman, 2012).

4.5 INFLUENCE OF BIOMARKER IN DIAGNOSTICS

The discovery of new biomarkers is critical for early illness identification. Though several age-related neurological illnesses are not inherited, some genetic factors have been established. Polygenic risk scores (PRSs) obtained from GWAS have shown potential in stratifying at-risk people across illness phases, as well as effectively identifying new genes that can be exploited as novel biomarkers and therapy targets (Rossi et al, 2023).

The APOEe4 mutation stands out as the most significant risk factor for AD, which leads in dysregulation of cholesterol; however, risk varies between ancestral upbringings and disorders. There are presently just a few diagnostic assays available to assess amyloid ratios and APOE status. Other indicators being studied concurrently include neurofilament proteins, phospho-tau species, and inflammatory markers (Raulin et al, 2022). Cerebrospinal fluid biomarkers are widely used in the diagnosis of dementia illnesses. Cerebrospinal fluid biomarkers such as total-tau, A-42, A-40, Neurofilament light chain, and phosphorylated-tau may help identify Frontotemporal lobar degeneration (FTLD) neuropathology and neurodegenerative-related alterations in vivo, aiding differential diagnosis (Virgilio et al, 2022).

Neurofilament light chains are fascinating possibilities in the extrapolative workup of neurodegeneration. Although the outcome of the Neurofilament light chain is moderate in healthy aging brains, it is accelerated in many disease-associated behaviours in multiple neurodegenerative disorders. Neurofilament light chain is widely employed in multiple sclerosis, numerous dementia illnesses such as AD, and several other NDDs such ALS (Vacchiano et al, 2021).

The discovery of accurate PD biomarkers may occur as a result of advances in neuroimaging methods such as, single-photon emission CT (SPECT), newer Magnetic Resonance Imaging (MRI) techniques, positron emission tomography (PET), and the advancement of new biochemical procedures. These strategies include measuring aggregated alpha-synuclein using protein misfolding cyclic amplification (PMCA), real-time quaking-induced conversion (RT-QuIC), and developing epigenetic-based biomarkers. Furthermore, combining biomarkers appears to be the most promising way to enhance their value and ensure a true finding of premotor PD (Delenclos et al, 2016).

4.6 MULTI-OMICS IN NEURODEGENERATION

Consequently, categorizing individuals into meaningful disease subgroups with clinical and biological significance could improve disease management and aid in the development of treatments. To that aim, omics skills such as proteomics, transcriptomics, genomics, and metabolomics are helping to provide a more complete understanding of the molecular processes implicit in the growth of NDs, allowing patients to be classified into subtypes founded on distinct molecular fingerprints (Figure 4.3).

4.6.1 For AD Patients

Omics analyses have significantly enhanced our comprehension of Alzheimer's pathogenesis. This progress is not only attributed to the revelation of its overall structures but also to the intricate detailing of circuits involving intricate molecular relations and regulations within crucial regions of brain affected by AD (Clark et al, 2021; De Jager et al, 2018; Nativio et al, 2020; Xicota et al, 2019; Zhou et al, 2021). An integrated approach to utilize proteomic, transcriptomic, and epigenomic analyses of autopsied human brains was employed to recognize molecular pathways associated with AD (Bai et al, 2020). Additionally, other investigations utilized an integrative approach involving blood RNAseq, lipidomics, and plasma metabolomics to establish a signature of blood-related omics predicting amyloid positivity in symptomless individuals who have risk (Mueller et al, 2005). This approach offers a less invasive, more accessible, and cost-effective means of assessing the risk of developing AD. A current study employing deep summarizing of the whole transcriptome, proteome, and phosphoproteome at different AD stages underscored the complementarity of deep proteomics technologies with genomics and transcriptomics. This study revealed vital molecular pathways and networks linked with AD, emphasizing that protein variations only moderately align with consistent RNA levels (Bennett et al, 2016). Collectively, these omics investigations, facilitated by the incorporation of openly existing datasets from international developments like the ROSMAP, and the ADNI serve as a valued source for

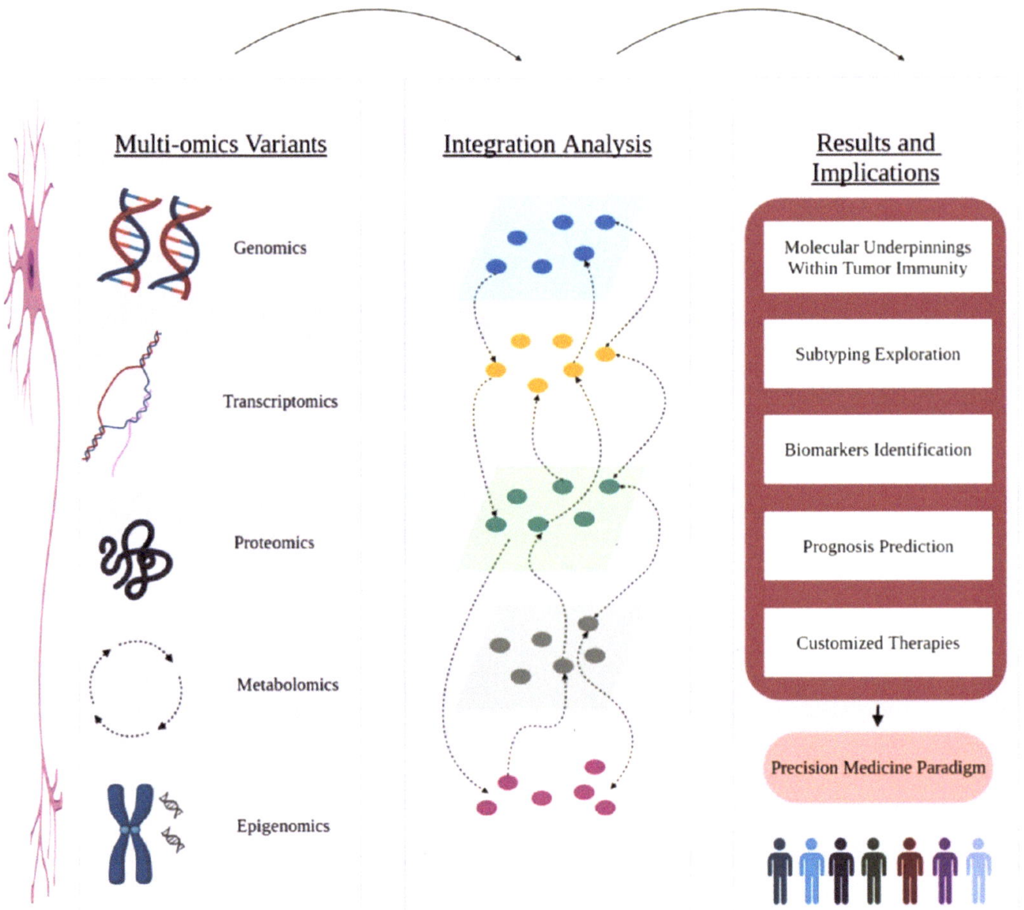

FIGURE 4.3 Multi-omics influence in neurology.

inclusive AD analyses. They also represent a possible catalyst for the discovery of novel biomarkers (A Bennett et al, 2012).

Beyond genomics, various omics fields such as transcriptomics, proteomics, and others like epigenomics and metabolomics contribute to a more precise delineation and understanding of Alzheimer's heterogeneity. In a recent study, conducted methylome-wide overtone analyses on brain and blood tissue trials from AD patients. They recognized methylation measurable attribute loci specific to certain groups, signifying a latent protagonist of epigenetic alterations in contributing to the diverse nature of Alzheimer's (Nazarian et al, 2020). Likewise, the prominent risk factor for irregular late-onset Alzheimer's, lipid chaperone protein, and APOE underscores the significance of investigating metabolic and lipid dynamics in AD research. Ongoing international projects like HUSERMET and Predict AD aim to define metabolic markers derived from serum in AD (Orešič et al, 2010). The outcomes of these metabolomics studies, along with others, enable the evaluation of various mixtures of metabolites for distinguishing diverse AD subtypes considered by distinct molecular mechanistic approach and clinical appearances. This progress contributes to the advancement of exertions to biochemically define heterogeneousness among AD patients (Badhwar et al, 2020; Bredesen, 2015; Czech et al, 2012; Wood et al, 2015).

4.6.2 For PD Patients

A recent intricate clinical trial has shed light on an intriguing aspect gleaned from omics data. This trial focuses on assessing the outcome of coenzyme Q10 in four distinct subgroups of PD patients categorized based on their omics-score predicting "mitochondrial risk burden." These subgroups include carriers of homozygous, heterozygous or Parkin/PINK1 mutations, patients with negative omics scores, and "omics" positive patients (Prasuhn et al, 2019). The trial aims to integrate data from those stratified groups, covering both non-motor symptoms and motor symptoms, magnetic resonance imaging results, and changes in brain structure.

Anticipated outcomes from this clinical trial could shed light on the effectiveness of the omics-score for PD patient stratification and offer insights into the potential for personalized treatment choices based on clinical, genetic, and neuroimaging data (Prasuhn et al, 2019). PM in PD, concentrating on mechanistically anchored disorders subgroups identified through combined omics results, holds aptitude (Rosen and Zeger, 2019). For instance, detailed peripheral tissues phenotypes in intermittent PD patients, whole-transcriptome RNA-seq pathway and compounding cellular assays and study along with genotyping data, allowed for patient lamination based on lysosomal (lyso-sPD) or mitochondrial (mito-sPD) primary disfunctions. This approach enabled the assortment of potential neuroprotective composites (Carling et al, 2020). This effective approach of merging by clinical phenotyping and inclusive analysis of transcriptomic, biological and genetic data, alongside an attentive evaluation of supposed neuroprotective compounds, presents a hopeful avenue for disease delamination and PM in infrequent PD.

Additional illustration is the Luxembourg Parkinson's project, an extensive molecular, clinical, and device-based ingenuity for identifying primary analysis and development markers in patients with atypical and typical parkinsonism (Hipp et al, 2018). A continuous, multi-site, and impartial investigation involves a comprehensive longitudinal assessment that includes omics-based molecular profiling studies using advanced biosamples. These biosamples encompass transcriptomics, metabolomics, proteomics, and genomics profiles obtained from saliva, blood, skin, cerebrospinal fluid (CSF), and urine. The study also incorporates a device-based assessment through a digital stage to complement international Parkinson cohort studies. This comprehensive and ongoing approach, ranging from intricate molecular analyses and genetic markers to continuous medical evaluations, aims to streamline the identification of PD subtypes, disease-specific biomarkers, and advancements in PM (Kia et al, 2021).

Kia et al. combined multiple bioinformatics tools, incorporating GWAS, TWAS, and methylation data, to pinpoint 11 potential genes associated with PD risk due to regulatory changes in expression, splicing, or methylation (Kia et al, 2021).

Additionally, protein level analyses and co-expression confirmed substantial functional relations with established genes of PD. The ongoing focus on integrating multi-level omics data, alongside advancements in understanding PD's underlying mechanisms, is set to improve current classification systems and biomarkers. This will enable personalized treatments and the development of more efficient therapeutic strategies.

4.6.3 For ALS Patients

Spinal cord and motor cortex samples of two distinct molecularly disconnected subgroups of ALS exhibited differential expression of a significant number of genes encrypting merging factors. Notably, SALS1 showed substantial overexpression, while SALS2 demonstrated down-regulation. Concurrently, analysis of number variants in the affected role revealed subtype-precise genomic modifications that linked positively with transcriptional signature profiles. This suggests a potential interplay between genomic and transcriptomic alterations contributing to the molecular heterogeneousness of basic ALS pathogenesis (Morello et al, 2019).

Moreover, a study indicated a correlation between low urate plasma levels and an elevated risk of emergent ALS, occurring several days before symptoms start (Bjornevik et al, 2019). Another study classified the metabolic profile of fibroblasts derived from skin samples of patients into specific subgroups based on their unique metabolic profiles. One subgroup exhibited amplified cysteine derived by trans-sulphuration to provision glucose hypermetabolism and glutathione biosynthesis distinguishing it from controls and other groups. The delamination of patients based on these molecular portraits lays the groundwork for the development of more effective and personalized therapeutic interventions for ALS. Our research revealed important biomarkers and targets that are irregularly regulated in certain ALS subgroups. This suggests their potential usefulness in creating customized medications based on specific molecular characteristics (Morello et al, 2015a; Morello and Cavallaro, 2015b). For instance, we observed dysregulation of histamine-related genes at the genomic and transcriptomic levels of two distinct subgroups of sporadic ALS (SALS) identified by molecular criteria. This supports the notion that histamine-allied target genes could serve as contender biomarkers and therapeutic goals for personalized ALS care. Opening results indicate that pharmacological variation of this signalling pathway appears to improve sclerosis features, enhancing motor survival and performance in ALS mice, as well as growing motor nerve cell existence in ALS models (Apolloni et al, 2017; Apolloni et al, 2019; Volonté et al, 2020; Bhandari and Kuhad, 2018).

4.7 ZONE OF LIMITATIONS

The progress of PM is confronted with several challenges that necessitate the integration of effective analytical utensils, databases, technologies, and methods to manage extensive and diverse data efficiently. This involves enhancing the collaboration and compatibility of public health, clinical systems, and laboratory. A significant obstacle to the application of PM is the demand for advanced data study. Presently, many PMI rely on semi-automated or manual processes, which are laborious and lack the capacity for on-demand study of varied human datasets. This limitation affects the ability to influence critical action spaces and predict possible ailment risks (Ahmed et al, 2020; Beltran et al, 2015; Lee et al, 2018; Van Panhuis et al, 2014; Luo et al, 2017; Katsanis, 2016; Manrai et al, 2016; Boyle et al, 2017; Shieh et al, 2017).

The conventional approach to computational examination involves successively a series of command-line applications, requiring proficient program design skills and familiarity with the UNIX environment. This method impedes the linkage of data produced during different stages of actions and tests conducted at various levels of sampling and sequencing. PM analyses demand intricate matched efforts among diverse category sets dealing with ununited data arrangements and substantial computation time, important for positively influencing treatment outcomes. Additionally, addressing social and ethical issues linked to healthcare facts collection, protection and privacy, is challenging but necessary (Bradford et al, 2014; Feero, 2014). The effective utilization of data science, information technology, and artificial intelligence (AI) has the potential to enhance public health research and monitoring. This entails the consistent gathering, organization, interpretation, and analysis of data within a shortened timeframe (Ahmed et al, 2020; Khoury et al, 2016; Khoury and Ioannidis, 2014; Bali et al, 2019; Buch et al, 2018). AI can play a crucial role in identifying metabolite penetrance by analyzing listed features and abnormalities, as well as assessing biochemical pathways in metabolites (Ahmed et al, 2013; Ahmed et al, 2014; Dandekar et al, 2014). This includes recognizing patterns of multimodal distributions associated with candidate genes (Ahmed et al, 2021; Tonn et al, 2019).

4.8 NEW FRANGLED AREA OF AI IN NEURODEGENERATION

AI has several potential applications in the treatment of neurodegenerative illnesses. For starters, it might be utilized to deliver more rapid and accurate diagnostics. Second, it can assist scientists in identifying possible pharmacological and biological targets that will lead to improved therapies. Finally, AI may be used to help scientists better understand the underlying causes of these problems, ultimately leading to potential therapies for a variety of neurological illnesses. This last issue is related to the recent emergence of personalized medicine, which is a type of treatment based on our genetic and molecular make-up. As AI advances, it is expected to be able to assist in the delivery of targeted medicines that are personalized to each patient. The medical industry is poised to achieve significant advances thanks to the usage of AI technologies. Much progress has been made, such as the continued growth of decision provision algorithms that assess pattern-based data, such as magnetic resonance imaging (MRI), X-rays, stained histology samples, and computed tomography (CT). AI is now being used to find and unravel increasingly complicated patterns, which will result in progress in discovering new treatment options and improving prognosis. AI shows significant potential as a tool to expand our understanding of complex NDDs in both current and future iterations (Stefano, 2023; Tăuţan et al, 2021).

AI is being used to review and detect trends in a variety of huge datasets, psychological tests counting co-morbidities, laboratory findings, lifestyle concerns, and neuroimaging as well as metagenomic and proteome data acquired from AD patients. Among the key objectives, researchers want to find designs that can be utilized to examine patients with moderate cognitive impairment (MCI) who eventually progress to Alzheimer's disease (AD; McKenzie et al, 2022).

An innovative study used AI techniques to detect nocturnal breathing issues as an objective marker for PD. Respiratory symptoms may appear and be detectable many years before visible motor signs (Armstrong and Okun, 2020; D'Arrigo et al, 2020). One finding of this AI-driven research is that PD can be predicted and tracked based on individual night-time breathing patterns, using formal sleep studies and RF sensors. This not only leads to early analysis but also informs investigators about new pathways and targets that can be used to recover current PD treatment strategies (Tăuţan et al., 2021).

A recent review looked at the use of AI in MS diagnosis, outcome prediction, and treatment monitoring based on laboratory findings and MRI (i.e., active mediators in blood and CSF). Among the challenges to overcome is that ALS is a complex illness that can be confused with numerous other conditions, both non-inflammatory and inflammatory, that exhibit similar clinical symptoms. Moreover, diagnosing the disease relies on observing how it spreads over time and space. Therefore, while AI methods improve disease diagnosis and care precision, expert supervision and assessment of all results remain crucial at present (Bonacchi et al, 2022).

4.9 FUTURE PERSPECTIVE AND CONCLUSION

The application of PM in neurology can be envisioned through two primary phases. The first phase involves the analysis of large-scale populations, encompassing all applicable categorizing variable quantity such as specific genomic and genetic makeup, diverse ethnicities, and sexes. This comprehensive approach considers the complex genetic and biological differences associated with different subgroups, enabling the identification of reliable molecular features and enough pathophysiological harmonies. Subsequently, each subsection can be embattled with actual therapeutic and defensive interferences. As AI-assisted medication blueprints become gradually harmonized in healthcare practice and clinical research, PM can progress to its second segment, characterized by truly adapted treatments. To achieve these distinguished goals, it is essential to recognize and embrace human assortment, warranting inclusion throughout the various phases of PM growth. This transformative journey aims to enhance health duration and improve treatments across an extensive range of disease situations, all within a broader outline dedicated to achieving brain health.

From a scientific perspective, the analysis of individual genomes paves the way for a novel era of defensive and modified medicine in healthcare. The rise of gene-targeted pharmaceuticals, precise targeting of molecular patterns in tumours, customized drug treatments, the ability to predict individual susceptibility to diseases, and advancements in the treatment and diagnosis of mental illnesses are among the transformative changes predictable in the coming decade. PM will empower doctors to seamlessly integrate health-related data with tests and targeted assays allowing for the identification and assessment of disease biomarkers and risks. This holistic approach enables the determination of actionable genetic alternatives in patients, a comprehensive understanding of the metabolome, and the mapping of metabolites to various pathway of diseases.

REFERENCES

A Bennett, D., et al., *Overview and findings from the religious orders study.* Current Alzheimer Research, 2012. **9**(6): p. 628–645.

Ahmed, Z., et al., *'Isotopo'a database application for facile analysis and management of mass isotopomer data.* Database, 2014. **2014**: p. bau077.

Ahmed, Z., et al., *Artificial intelligence with multi-functional machine learning platform development for better healthcare and precision medicine.* Database, 2020. **2020**: p. baaa010.

Ahmed, Z., et al., *Integrative clinical, genomics and metabolomics data analysis for mainstream precision medicine to investigate COVID-19.* BMJ Innovations, 2021. **7**(1): 1–10.

Ahmed, Z., et al., *Software LS-MIDA for efficient mass isotopomer distribution analysis in metabolic modelling.* BMC Bioinformatics, 2013. **14**: p. 1–11.

Apolloni, S., et al., *Histamine regulates the inflammatory profile of SOD1-G93A microglia and the histaminergic system is dysregulated in amyotrophic lateral sclerosis.* Frontiers in Immunology, 2017. **8**: p. 1689.

Apolloni, S., et al., *Histaminergic transmission slows progression of amyotrophic lateral sclerosis.* Journal of Cachexia, Sarcopenia and Muscle, 2019. **10**(4): p. 872–893.

Armstrong, M.J. and M.S. Okun. Diagnosis and treatment of Parkinson disease: A review. JAMA, 2020. 323: p. 548–560. doi: 10.1001/jama.2019.22360.

Badhwar, A., et al., *A multiomics approach to heterogeneity in Alzheimer's disease: Focused review and roadmap.* Brain, 2020. **143**(5): p. 1315–1331.

Bai, B., et al., *Deep multilayer brain proteomics identifies molecular networks in Alzheimer's disease progression.* Neuron, 2020. **105**(6): p. 975–991. e7.

Bali, J., R. Garg and R.T. Bali, *Artificial intelligence (AI) in healthcare and biomedical research: Why a strong computational/AI bioethics framework is required?* Indian Journal of Ophthalmology, 2019. **67**(1): p. 3.

Bartels, A.L. and K.L. Leenders, *Parkinson's disease: The syndrome, the pathogenesis and pathophysiology*. Cortex, 2009. **45**(8): p. 915–921.

Bayer, T.A. and O. Wirths, *Intracellular accumulation of amyloid-Beta - A predictor for synaptic dysfunction and neuron loss in Alzheimer's disease*. Frontiers in Aging Neuroscience, 2010. **2**: p. 8.

Beltran, H., et al., *Whole-exome sequencing of metastatic cancer and biomarkers of treatment response*. JAMA Oncology, 2015. **1**(4): p. 466–474.

Bhandari, R. and A. Kuhad, *Edaravone: A new hope for deadly amyotrophic lateral sclerosis*. Drugs of Today (Barcelona, Spain: 1998), 2018. **54**(6): p. 349–360.

Bjornevik, K., et al., *Prediagnostic plasma metabolomics and the risk of amyotrophic lateral sclerosis*. Neurology, 2019. **92**(18): p. e2089–e2100.

Bonacchi, R., M. Filippi and M.A. Rocca. Role of artificial intelligence in MS clinical practice. NeuroImage Clinical, 2022. 35: 103065. doi: 10.1016/j.nicl.2022.103065.

Boyle, E.A., Y.I. Li and J.K. Pritchard, *An expanded view of complex traits: From polygenic to omnigenic*. Cell, 2017. **169**(7): p. 1177–1186.

Bradford, W., et al., *Development of a HIPAA-compliant environment for translational research data and analytics*. Journal of the American Medical Informatics Association, 2014. **21**(1): p. 185–189.

Bredesen, D.E., *Metabolic profiling distinguishes three subtypes of Alzheimer's disease*. Aging (Albany NY), 2015. **7**(8): p. 595.

Buch, V.H., I. Ahmed and M. Maruthappu, *Artificial intelligence in medicine: Current trends and future possibilities*. British Journal of General Practice, 2018. **68**(668): p. 143–144.

Butterfield, D.A., et al., *Amyloid beta-peptide and amyloid pathology are central to the oxidative stress and inflammatory cascades under which Alzheimer's disease brain exists*. Journal of Alzheimer's Disease, 2002. **4**(3): p. 193–201.

Cacace, R., et al., *Rare variants in PLD3 do not affect risk for early-onset Alzheimer disease in a European consortium cohort*. Human Mutation, 2015. **36**(12): p. 1226–1235.

Carling, P.J., et al., *Deep phenotyping of peripheral tissue facilitates mechanistic disease stratification in sporadic Parkinson's disease*. Progress in Neurobiology, 2020. **187**: p. 101772.

Clark, C., et al., *An integrative multi-omics approach reveals new central nervous system pathway alterations in Alzheimer's disease*. Alzheimer's Research & Therapy, 2021. **13**(1): p. 1–19.

Collins, F.S. and H. Varmus, *A new initiative on precision medicine*. The New England Journal of Medicine, 2015. **372**(9): p. 793–795.

Czech, C., et al., *Metabolite profiling of Alzheimer's disease cerebrospinal fluid*. PLoS One, 2012. **7**(2): p. e31501.

D'Arrigo, A., S. Floro, F. Bartesaghi, C. Casellato, G. Francesco, S. Papa, S. Centanni, A. Priori and T. Bocci. Respiratory dysfunction in Parkinson's disease: A narrative review. ERJ Open Research, 2020. 6: p. 00165–2020. doi: 10.1183/23120541.00165-2020.

Dandekar, T., et al., *Software applications toward quantitative metabolic flux analysis and modeling*. Briefings in Bioinformatics, 2014. **15**(1): p. 91–107.

De Jager, P.L., et al., *A multi-omic atlas of the human frontal cortex for aging and Alzheimer's disease research*. Scientific Data, 2018. **5**(1): p. 1–13.

de Paula, V.J.R., et al., *Neurobiological pathways to Alzheimer's disease: Amyloid-beta, TAU protein or both?* Dementia & Neuropsychologia, 2009. **3**(3): p. 188–194.

Delenclos, M., et al., *Biomarkers in Parkinson's disease: Advances and strategies*. Parkinsonism & Related Disorders, 2016. **22**(Suppl 1): p. S106–S110.

Emamzadeh, F.N. and A. Surguchov, *Parkinson's disease: Biomarkers, treatment, and risk factors*. Frontiers in Neuroscience, 2018. **12**: p. 612.

EpiPM Consortium, *A roadmap for precision medicine in the epilepsies*. Lancet Neurology, 2015. **14**(12): p. 1219–1228.

Feero, W.G., *Clinical application of whole-genome sequencing: Proceed with care.* JAMA, 2014. **311**(10): p. 1017–1019.

Gorell, J.M., et al., *Multiple risk factors for Parkinson's disease.* Journal of the Neurological Sciences, 2004. **217**(2): p. 169–174.

Green, E.D., et al., *Strategic vision for improving human health at the forefront of genomics.* Nature, 2020. **586**(7831): p. 683–692.

Hardiman, O., et al., *Amyotrophic lateral sclerosis.* Nature Reviews Disease Primers, 2017. **3**: p. 17085.

Harold, D., et al., *Genome-wide association study identifies variants at CLU and PICALM associated with Alzheimer's disease.* Nature Genetics, 2009. **41**(10): p. 1088–1093.

Hess, J.L., et al., *A polygenic resilience score moderates the genetic risk for schizophrenia.* Molecular Psychiatry, 2021. **26**(3): p. 800–815.

Hipp, G., et al., *The Luxembourg Parkinson's study: A comprehensive approach for stratification and early diagnosis.* Frontiers in Aging Neuroscience, 2018. **10**: p. 326.

Hood, L. and S.H. Friend, *Predictive, personalized, preventive, participatory (P4) cancer medicine.* Nature Reviews Clinical Oncology, 2011. **8**(3): p. 184–187.

Jiang, T., et al., *Genetic variation in PICALM and Alzheimer's disease risk in Han Chinese.* Neurobiology of Aging, 2014. **35**(4): p. 934.e1-3.

Katsanis, N., *The continuum of causality in human genetic disorders.* Genome Biology, 2016. **17**: p. 1–5.

Kesselheim, A.S., T.J. Hwang and J.M. Franklin, *Two decades of new drug development for central nervous system disorders.* Nature Reviews Drug Discovery, 2015. **14**(12): p. 815–816.

Khoury, M.J., M.F. Iademarco and W.T. Riley, *Precision public health for the era of precision medicine.* American Journal of Preventive Medicine, 2016. **50**(3): p. 398–401.

Khoury, M.J. and J.P. Ioannidis, *Big data meets public health.* Science, 2014. **346**(6213): p. 1054–1055.

Kia, D.A., et al., *Identification of candidate Parkinson disease genes by integrating genome-wide association study, expression, and epigenetic data sets.* JAMA Neurology, 2021. **78**(4): p. 464–472.

Kouli, A., K.M. Torsney and W.L. Kuan, *Parkinson's Disease: Etiology, Neuropathology, and Pathogenesis, in Parkinson's Disease: Pathogenesis and Clinical Aspects*, T.B. Stoker and J.C. Greenland, Editors. 2018, Codon Publications Copyright: The Authors: Brisbane (AU).

Kuopio, A.M., et al., *Environmental risk factors in Parkinson's disease.* Movement Disorders, 1999. **14**(6): p. 928–939.

Lamptey, R.N.L., et al., *A review of the common neurodegenerative disorders: Current therapeutic approaches and the potential role of nanotherapeutics.* International Journal of Molecular Sciences, 2022. **23**(3): p. 1851.

Lee, S.-I., et al., *A machine learning approach to integrate big data for precision medicine in acute myeloid leukemia.* Nature Communications, 2018. **9**(1): p. 42.

Luo, Y., F.S. Ahmad and S.J. Shah, *Tensor factorization for precision medicine in heart failure with preserved ejection fraction.* Journal of Cardiovascular Translational Research, 2017. **10**: p. 305–312.

Ma, X.Y., et al., *Missense variants in CR1 are associated with increased risk of Alzheimer' disease in Han Chinese.* Neurobiology of Aging, 2014. **35**(2): p. 443.e17-21.

Manrai, A.K., J.P. Ioannidis and I.S. Kohane, *Clinical genomics: From pathogenicity claims to quantitative risk estimates.* JAMA, 2016. **315**(12): p. 1233–1234.

Mazzio, E.A. and K.F. Soliman, *Basic concepts of epigenetics: Impact of environmental signals on gene expression.* Epigenetics, 2012. **7**(2): p. 119–130.

McKenzie, A.T., G.A. Marx, D. Koenigsberg, M. Sawyer, M.A. Iida, J.M. Walker, T.E. Richardson, G. Campanella, J. Attems and A.C. McKee, et al. *Interpretable deep learning of myelin histopathology in age-related cognitive impairment.* Acta Neuropathologica Communications, 2022. 10: p. 131. doi: 10.1186/s40478-022-01425-5.

Mondragón-Rodríguez, S., et al., *Amyloid Beta and tau proteins as therapeutic targets for Alzheimer's disease treatment: Rethinking the current strategy.* International Journal of Alzheimer's Disease, 2012. **2012**: p. 630182.

Morello, G., et al., *Integrative multi-omic analysis identifies new drivers and pathways in molecularly distinct subtypes of ALS.* Scientific Reports, 2019. **9**(1): p. 9968.

Morello, G., et al., *Selection of potential pharmacological targets in ALS based on whole-genome expression profiling.* Current Medicinal Chemistry, 2015a. **22**(17): p. 2004–2021.

Morello, G. and S. Cavallaro, *Transcriptional analysis reveals distinct subtypes in amyotrophic lateral sclerosis: Implications for personalized therapy.* Future Medicinal Chemistry, 2015b. **7**(10): p. 1335–1359.

Morris, J., *Amyotrophic lateral sclerosis (ALS) and related motor neuron diseases: An overview.* The Neurodiagnostic Journal, 2015. **55**(3): p. 180–194.

Mueller, S.G., et al., *The Alzheimer's disease neuroimaging initiative.* Neuroimaging Clinics, 2005. **15**(4): p. 869–877.

Nativio, R., et al., *An integrated multi-omics approach identifies epigenetic alterations associated with Alzheimer's disease.* Nature Genetics, 2020. **52**(10): p. 1024–1035.

Nazarian, A., A.I. Yashin and A.M. Kulminski, *Summary-based methylome-wide association analyses suggest potential genetically driven epigenetic heterogeneity of Alzheimer's disease.* Journal of Clinical Medicine, 2020. **9**(5): p. 1489.

Orešič, M., J. Lötjönen and H. Soininen, *Systems medicine and the integration of bioinformatic tools for the diagnosis of Alzheimer's disease.* Genome Medicine, 2010. **2**(11): p. 1–5.

Pickett, E.K., et al., *Amyloid Beta and tau cooperate to cause reversible behavioral and transcriptional deficits in a model of Alzheimer's disease.* Cell Reports, 2019. **29**(11): p. 3592–3604.e5.

Prasuhn, J., et al., *An omics-based strategy using coenzyme Q10 in patients with Parkinson's disease: Concept evaluation in a double-blind randomized placebo-controlled parallel group trial.* Neurological Research and Practice, 2019. **1**(1): p. 1–7.

Priyadarshi, A., et al., *Environmental risk factors and Parkinson's disease: A metaanalysis.* Environmental Research, 2001. **86**(2): p. 122–127.

Raulin, A.C., et al., *ApoE in Alzheimer's disease: Pathophysiology and therapeutic strategies.* Molecular Neurodegeneration, 2022. **17**(1): p. 72.

Rosen, A. and S.L. Zeger, *Precision medicine: Discovering clinically relevant and mechanistically anchored disease subgroups at scale.* The Journal of Clinical Investigation, 2019. **129**(3): p. 944–945.

Rossi, S.L., P. Subramanian and D.E. Bovenkamp, *The future is precision medicine-guided diagnoses, preventions and treatments for neurodegenerative diseases.* Frontiers in Aging Neuroscience, 2023. **15**: 1128619.

Shieh, Y., et al., *Breast cancer screening in the precision medicine era: Risk-based screening in a population-based trial.* Journal of the National Cancer Institute, 2017. **109**(5): p. djw290.

Spires-Jones, T.L., J. Attems and D.R. Thal, *Interactions of pathological proteins in neurodegenerative diseases.* Acta Neuropathologica, 2017. **134**(2): p. 187–205.

Spires-Jones, T.L. and B.T. Hyman, *The intersection of amyloid beta and tau at synapses in Alzheimer's disease.* Neuron, 2014. **82**(4): p. 756–771.

Stefano, G.B. *Artificial intelligence as a tool for the diagnosis and treatment of neurodegenerative diseases.* Brain Science, 2023. **13**(6): 938. doi: 10.3390/brainsci13060938.

Tan, L., et al., *Genome-wide serum microRNA expression profiling identifies serum biomarkers for Alzheimer's disease.* Journal of Alzheimer's Disease, 2014a. **40**(4): p. 1017–1027.

Tan, M.S., et al., *Genetic variation in BIN1 gene and Alzheimer's disease risk in Han Chinese individuals.* Neurobiology of Aging, 2014b. **35**(7): p. 1781.e1-8.

Tăuţan, A.M., B. Ionescu and E. Santarnecchi, *Artificial intelligence in neurodegenerative diseases: A review of available tools with a focus on machine learning techniques.* Artificial Intelligence in Medicine, 2021. **117**: p. 102081.

Tonn, M.K., et al., *Stochastic modelling reveals mechanisms of metabolic heterogeneity.* Communications Biology, 2019. **2**(1): p. 108.

Vacchiano, V., et al., *Plasma and CSF neurofilament light chain in amyotrophic lateral sclerosis: A cross-sectional and longitudinal study*. Frontiers in Aging Neuroscience, 2021. **13**: p. 753242.

Van Panhuis, W.G., et al., *A systematic review of barriers to data sharing in public health*. BMC Public Health, 2014. **14**(1): p. 1–9.

Vardarajan, B.N., et al., *Rare coding mutations identified by sequencing of Alzheimer disease genome-wide association studies loci*. Annals of Neurology, 2015. **78**(3): p. 487–498.

Virgilio, E., et al., *The role of tau beyond Alzheimer's disease: A narrative review*. Biomedicines, 2022. **10**(4): p. 760.

Volonté, C., et al., *Omics-based exploration and functional validation of neurotrophic factors and histamine as therapeutic targets in ALS*. Ageing Research Reviews, 2020. **62**: p. 101121.

Wang, G.Y., et al., *Advances in nanotechnology-based strategies for the treatments of amyotrophic lateral sclerosis*. Materials Today Bio, 2020. **6**: p. 100055.

Wood, P.L., et al., *Targeted lipidomics of frontal cortex and plasma diacylglycerols (DAG) in mild cognitive impairment and Alzheimer's disease: Validation of DAG accumulation early in the pathophysiology of Alzheimer's disease*. Journal of Alzheimer's Disease, 2015. **48**(2): p. 537–546.

Xicota, L., et al., *Multi-omics signature of brain amyloid deposition in asymptomatic individuals at-risk for Alzheimer's disease: The INSIGHT-preAD study*. EBioMedicine, 2019. **47**: p. 518–528.

Yu, M., O. Sporns and A.J. Saykin, *The human connectome in Alzheimer disease - relationship to biomarkers and genetics*. Nature Reviews Neuroscience, 2021. **17**(9): p. 545–563.

Zhou, Y., et al., *AlzGPS: A genome-wide positioning systems platform to catalyze multi-omics for Alzheimer's drug discovery*. Alzheimer's Research & Therapy, 2021. **13**(1): p. 1–13.

5 Sleep Deprivation and Cognitive-Behavioral Impairment in Neurodegenerative Disorders

Adeel Ahmed, Mushahid Hussain, Niaz Ahmed, Abid Ali, Farnaz Norouziyan, Muhammad Umar, and Nasrollah Moradikor

5.1 INTRODUCTION

Chronic sleep deprivation or disturbance is a characteristic feature of modern society (Meerlo et al., 2015). People who are constantly busy and under a lot of stress often don't get enough sleep. This is due to the fact that several individuals compromise their sleep in order to engage in employment and generate additional income. Neuronal repair heavily relies on sleep, which reduces the burden on brain plasticity. Therefore, sleep deprivation (SD) can significantly impact neurological functioning (Born and Feld, 2012). In addition, SD has led to a pervasive health catastrophe among the populace, resulting in substantial repercussions on health, economy, and society (Hafner et al., 2017). Hence, it is crucial to examine the brain mechanisms that are responsive to SD to devise specific treatments. SD is currently recognized as a prominent potential determinant of a variety of risks of neuropsychiatric diseases, including depression (Roberts and Duong, 2014). Most people with depression commonly have disruptions in their sleep patterns (Ding et al., 2019; Mahato and Paul, 2019). Depression and anxiety are concurrent conditions (Cryan and Holmes, 2005). Various epidemiological studies have demonstrated that insistent sleep disruptions have a role in the onset of chronic anxiety and depression (Chapman et al., 2013; Roberts and Duong, 2014). Many research studies have confirmed that chronic sleep problems gradually cause alterations in the brain that closely look like those seen in individuals with depression (Buysse et al., 2008; Meerlo et al., 2015). Recent evidence indicates that microglia, the main immune cells in the brain, have a substantial impact on the development and progression of depression (Jia et al., 2021). Furthermore, there have been documented instances of activated microglia in the brains of people suffering from depression, as well as in animal models (Maes et al., 2009; Beumer et al., 2012; Frick et al., 2013; Prinz and Priller, 2014). Activation of glial cells, which are the brain's natural immune cells, can induce an imbalance in cytokine levels, resulting in behavioural problems. Furthermore, previous research has demonstrated that sleep and SD have divergent effects on glial cells (Bellesi et al., 2017; Choudhury et al., 2020). The Keap1 and Nrf2 system, composed of the Kelch-like ECH-associated protein 1 and nuclear factor erythroid 2-related factor, exerts

DOI: 10.1201/9781032711102-5

a substantial influence on inflammation, which is connected with depression (Hashimoto, 2018). Nrf2 is a crucial transcription factor involved in protecting against harm induced by electrophilic and oxidative stresses (Ma, 2013). It attaches to the antioxidant response elements found in the gene promoters that are responsible for creating different enzymes and proteins involved in the process of detoxification or oxidation-reduction that react to stress. The role of Nrf2 in regulating oxidative stress and preventing toxicity has been discussed in various studies (Ma, 2013; Suzuki and Yamamoto, 2015). The Keap1-Nrf2 pathway governs the regulation of the function of several enzymes involved in cellular protection and detoxification. This pathway is crucial for preserving the equilibrium of cellular redox processes. Moreover, the function of Nrf2 is intricately regulated by genes linked to sleep, suggesting that sleep plays a role in diminishing oxidative stress and inflammation by modulating Nrf2 signaling (Wible et al., 2018; Spiers et al., 2019). Neuroinflammation leads to the emergence of anxiety and depression, whereas the administration of anti-inflammatory drugs can relieve these symptoms. Minocycline, a second-generation tetracycline, is being evaluated as a potential therapy for depression and anxiety because of its notable ability to reduce inflammation and protect the nervous system characteristics, as well as its ability to cross the blood-brain barrier (Pae et al., 2008; Soczynska et al., 2012). Minocycline achieves its neuroprotective effects by effectively suppressing the development and activation of microglia (Tikka et al., 2001; Kobayashi et al., 2013a). Multiple research investigations have demonstrated that administering minocycline effectively reduces the symptoms associated with anxiety and depression (Han et al., 2020). Prior study has also shown that minocycline exhibits a neuroprotective influence on spatial memory deficits and abnormalities in adult neurogenesis caused by SD (Wadhwa et al., 2017).

5.2 PHYSIOLOGICAL ROLE OF SLEEP

Sleep is a vital physiological activity that allows the body to relax and supports several brain activities, including mood regulation, memory consolidation, consciousness, energy restoration, and the creation of important macromolecules like growth hormones (Mackiewicz et al., 2007). Authors also include executive functioning (Slama et al., 2018), immuno-protection (Iliff et al., 2012), and the repair of neural damage (Zada et al., 2019). In the absence of external factors, sleep is controlled by two primary mechanisms: the homeostatic process and the circadian system. The circadian system regulates the timing of sleep-wake cycles and causes changes in hormone and neurotransmitter levels throughout the day. On the other side, the homeostatic regulation of sleep creates a build-up of sleep pressure as time awake increases, which decreases when one is sleeping. The circadian system also controls the timing of sleep-wake cycles and causes fluctuations in hormone and neurotransmitter levels on a daily basis. Under typical circumstances, the two systems collaborate to preserve the structural stability of sleep throughout the night and wakefulness during the day. Sleep is classified into two categories: non-rapid eye movement (NREM) sleep, also known as non-REMS, and rapid eye movement sleep (REMS). Typically, the initial transition into NREM sleep occurs while young adults are in a drowsy state. NREM sleep, often known as dreamless sleep, is categorized into four separate stages (Zammit et al., 1999; Huyett et al., 2021; Rogers et al., 2021). The initial phase, known as N1, is the least intense stage, marked by a shift from being awake to falling asleep. Next, the second stage N2 occurs, which accounts for the majority of the overall sleep duration. Subsequently, the adult transitions into profound stages of sleep known as N3 and N4, characterized by distinct electroencephalogram (EEG) waves. These phases follow a well-regulated pattern, alternating to form sleep cycles (Malhotra, 2018). EEG shows more low-frequency waves with high amplitudes during NREM sleep. The eyes do not move. NREM stage N3, also known as slow-wave sleep (SWS), occurs approximately 70–80 minutes into a deep sleep. During this stage, sleep becomes lighter and exhibits key features of REM sleep, such as heightened activity in the high-frequency range of the EEG and greater REM. REM sleep is the last stage of a sleep cycle, also known as paradoxical sleep. During this stage, the EEG activity resembles that of wakefulness, with low-voltage and higher-frequency signals. Additionally, the eyelids move rapidly while the muscles remain entirely relaxed. The REM period typically occurs during active dreaming and is characterized by physical movements. This process is performed periodically at intervals of around 90 minutes. During the early morning, the amount of deep stage 3 and 4 sleep decreases while the amount of REM sleep increases (Petit et al., 2004).

5.3 THE IMPACT OF SLEEP LOSS ON NEUROINFLAMMATION AND ACTIVATION OF MICROGLIA

SD triggers a multitude of inflammatory reactions (Atrooz and Salim, 2020). Insufficient sleep is a prevalent and substantial global health issue that is associated with the modulation of the immune system (Dworak et al., 2011; Allen and Lyons, 2018) and loss of mood (Babson et al., 2010; Alkadhi et al., 2013). Due to the diverse range of studies and differences in how sleep disruption is defined, it is challenging to make conclusions regarding the influence of sleep on inflammation. This is further complicated by the use of different methods to assess sleep disturbance (such as sleep quality and sleep grievances), variations in sleep period, and the use of many markers to measure inflammation (Irwin, 2015). SD poses a significant oxidative strain on the brain. Although it is well known that neuroinflammation and oxidative stress have significant roles in the detrimental consequences of sleep loss, our comprehension of the mechanisms that can mitigate these consequences is still restricted. Research indicates a significant occurrence of anxiety and sadness, along with a connection between inflammation and many medical disorders (Kang et al., 2011; Konsman, 2019). Microglial cells undergo commencement in response to an inflammatory stimulus, which influences the secretion of cytokines (pro-inflammatory cytokines rise while anti-inflammatory cytokines reduce), typically under observable behavioral changes (Kang et al., 2014; Wohleb et al., 2014). The evidence supporting the existence of bi-directional contact between the sleep and immune systems is increasing. Sleep and the immune system have a reciprocal effect on each other (Zielinski and Krueger, 2011). Glial cells are the intrinsic immune cells of the brain. When activated, these cells lead to abnormalities in cytokines, which in turn induce disruptions in behaviour. Nevertheless, the specific function they serve in sleep-deprived circumstances remains uncertain (Wisor et al., 2011; Alkadhi et al., 2013). Insufficient sleep has been found to impair mental performance and elevate the risk of immunological dysfunctions (Wisor et al., 2011). Pro-inflammatory cytokines are linked to both SD (sleep disorders) and mood disorders, but the exact mechanism behind this association is not well understood (Ronnback and Hansson, 2004; Abelaira et al., 2013; Hong et al., 2016). Microglia have a crucial role as immunological sensors in the brain. Prior research indicates that microglia detect stresses associated with depression and trigger immunological responses and neuroinflammation that contribute to the onset of depression. Although there has been some advancement in comprehending the function of microglia in depression, there are still some areas that require further investigation. Concepts continue to be difficult to grasp. Initially, it is crucial to ascertain, within a particular area or network, whether microglia are the primary agents that respond to stress or the secondary factors that come into play following stress-induced alterations in neurons. Furthermore, it is crucial to monitor the movements and changes in microglial cells. To obtain a comprehensive understanding of microglial phenotypes during the progression of depression, it is necessary to examine many brain regions at different phases and establish a reliable spatiotemporal profile. Additionally, there is a need for further mechanistic research to comprehend how stress-induced alterations in microglial cellular pathways contribute to the onset of depression. The subject of microglial heterogeneity in various brain areas has been addressed (Olah et al., 2011; Halaris, 2019).

5.4 THE INVOLVEMENT OF MICROGLIA IN DEPRESSION AND ANXIETY

The Latin term "glue", glia, serves not just to physically connect neurons but also to actively regulate neural functions and uphold brain equilibrium (Allen and Lyons, 2018). Astrocytes, oligodendrocytes, ependymal cells, and microglia constitute the predominant population of glial cells. Microglia constantly monitor the microenvironment using their extremely mobile and branched processes, which are classified according to their specific origin and morphology (refer to the following section) (Kettenmann et al., 2011). Microglia, the brain's predominant immune cells that serve as the initial defense mechanism, have garnered significant interest in recent years due to their role in facilitating immunological reactions and neuroinflammation (Wolf et al., 2017). Microglia spontaneously assume a branched shape in the absence of external triggers, with compact and immobile cell bodies but mobile extensions that constantly survey the brain tissue to uphold brain homeostasis (Kettenmann et al., 2013). Microglia exhibit significant morphological alterations in response to environmental stimuli (Kettenmann et al., 2013). Following injury or disease, ramified microglia undergo a fast transformation into an amoeboid morphology characterized by swelling soma and shorter processes. This transformation is accompanied by an increase in phagocytic activity and cytokine production. The term used to describe this process is microglial activation. Activated microglia show various reactions to external stimuli and function as a two-sided weapon: immediate microglial activation frequently aids in tissue restoration by eliminating intruding. Pathogens and cellular waste are cleared by microglia, but prolonged activation of microglia leads to persistent neuroinflammation, worsening damage, and promoting the progression of illness (Kettenmann et al., 2011). Accumulating data indicate that inflammation or stressors can cause dysfunctions in microglia, which are typically observed in cases of depression. These dysfunctions may have a causative role in disturbing brain functioning and ultimately contributing to depression (Yirmiya et al., 2015). Clinical research strongly suggests that inflammation-related traits have a significant role in depressed patients (Miller and Raison, 2016). Peripheral immune cells and microglia have been linked to depression-related inflammation (Wohleb et al., 2014; Miller and Raison, 2016). Prolonged elevated inflammation in the body is associated with a higher risk of illnesses; its impact on the brain can result in behavioural manifestations associated with mood and anxiety disorders, which are frequently coexisting conditions (Gorman, 1996). It is believed that the body's release of inflammatory signals influences the brain, contributing to behavioural symptoms associated with mood and anxiety disorders. Peripheral immune cells may enter the brain through this pathway, and cytokines may activate local inflammatory signaling pathways and microglia (D'mello et al., 2009; Wohleb et al., 2012).

5.5 Keap1-Nrf2 AND SLEEP DEPRIVATION

Sleep is an essential physiological function that is preserved in the human body. The amount and quality of sleep can have an impact on an individual's overall health and quality of life (Gao et al., 2019). Despite the recommendation for individuals to sleep for seven to nine hours per day, there is a significant occurrence of inadequate sleep in modern societies (Javaheripour et al., 2019). Extensive evidence supports the fact that insufficient sleep has negative effects on human well-being (McDonald and Keene, 2010; Mullin et al., 2013). Sleep deprivation (SD) has been widely recognized for its detrimental effects on neurobehavior, including improved oxidative stress and inflammation in the brain. The nuclear factor erythroid 2-related factor 2 (Nrf2) is a crucial regulator of the brain's internal protective mechanisms against oxidative stress. Typically, Nrf2 is found in the cytoplasm and interacts with a protein called Kelch-like ECH-associated protein 1 (Keap-1), which leads to its breakdown by proteasomes. However, when Nrf2 is activated, it moves into the nucleus to control the activity of its subsequent enzymes (Mcdonald and Keene, 2010). Concurrently, Nrf2 signaling also has a crucial function in regulating inflammatory responses. Multiple studies have shown that the Nrf2-modulated antioxidant system can effectively work together to control inflammation through diverse mechanisms (Ghosh et al., 2017; Xu et al., 2017).

5.6 Keap1-Nrf2 RESPONSE IN OXIDATIVE STRESS

The Nrf2 protein is confined by the Keapl protein in the cytoplasm and is consistently marked for polyubiquitination in a fundamental state. Exposure of cells to electrophilic and oxidative stress triggers the liberation of Nrf2 from Keapl, allowing Nrf2 to relocate. The nucleus attaches to the ARE and activates a battery of phase II cytoprotective genes. Nrf2 not only regulates phase II cytoprotective genes but also controls the transcription of different drug-metabolizing enzymes.

5.7 INFLAMMATION AND Keap1-Nrf2

Inflammation is a complicated physiological response that happens when tissues are contaminated or wounded by adverse stimuli, such as infections, irritants, or damage. This protective reaction involves the participation of immune cells, blood vessels, and molecular mediators (Katoh et al., 2001; Ferrero-Miliani et al., 2007; Farooqui et al., 2021). Inflammation is a pathological occurrence that is linked to several illness conditions, primarily caused by chemical, physical, biological, and psychological causes. The purpose of inflammation is to restrict and eradicate the factors that cause harm to cells, remove or absorb dead cells and tissues, and commence the process of tissue restoration. There are two different types of inflammation: acute and chronic. Acute inflammation is a temporary and advantageous response in the body, while chronic inflammation, which persists over a long period, is frequently observed in various chronic diseases and problems. Chronic inflammation is caused by the direct invasion of various mononuclear immune cells, including monocytes, macrophages, lymphocytes, and plasma cells, as well as the generation of inflammatory cytokines. Chronic inflammation is acknowledged. Keap1-Nrf2 has been demonstrated to have a significant impact on various inflammatory disorders impacting multiple systems in living organisms. These diseases encompass colitis, arthritis, gastritis, liver damage, pneumonia, cardiovascular disease, neurodegenerative disease, and brain damage. Investigations revealed that Nrf2-/- animals exhibited more pronounced symptoms of inflammation and tissue injury compared to wild-type (WT) animals. Hence, it is well accepted that the Nrf2 signaling pathway exerts a safeguarding influence on inflammatory disorders (Yu et al., 2014; Gore et al., 2016; Shi et al., 2018).

5.8 Keap1-Nrf2 IN DEPRESSION AND ANXIETY

Depression, a prevalent psychiatric condition globally, is a mood illness characterized by a significant likelihood of decline. According to the World Health Organization (WHO), about 320 million people of all age groups are affected by depression, making it a significant cause of the worldwide disease burden (Woody et al., 2017). While the particular mechanisms behind the development of depression are not yet understood, there is growing evidence suggesting that inflammatory processes play a role in its pathogenesis (Dantzer et al., 2008; Hashimoto, 2009, 2015; Miller and Raison, 2016; Zhang et al., 2016; Miller et al., 2017). The Keap1-Nrf2 plays an essential role in the pathophysiology of depression given its crucial involvement in inflammation, and the Keap1-Nrf2 system is responsible for regulating the cellular defense against oxidative stress, with Nrf2 acting as the main transcription factor. It attaches to the antioxidant response elements (ARE) found in the promoter region of genes that encode various phase II detoxifying or antioxidant enzymes and related stress-responsive proteins (Kobayashi et al., 2013a; Suzuki and Yamamoto, 2015; Yamamoto et al., 2018). In normal circumstances, the activity of Nrf2 is inhibited by Keap1, which serves as an adapter protein responsible for the breakdown of Nrf2 (Suzuki and Yamamoto, 2015). Under conditions of oxidative stress, the transcription factor Nrf2 is released from repression and initiates the activation of genes that provide protection (Suzuki et al., 2013; Suzuki and Yamamoto, 2015). The Keap1-Nrf2 system is significantly involved in the development of pathologies linked with inflammation (Suzuki et al., 2013; Kobayashi et al., 2013a; Suzuki and Yamamoto, 2015). The findings indicate that reduced Keap1-Nrf2 signaling is a significant factor in the development of mood disorders, including major depressive disorder (MDD) and bipolar disorder (Zhang et al., 2018). The Keap1-Nrf2 system is reduced due to inflammation or stress, leading to the development of depression-like symptoms (Yao et al., 2016). Psychiatric diseases in modern society often involve physiological reactions and altered behavioural states, such as anxiety-like behaviours, which are accompanied by feelings of nervousness and anxiety (Alonso et al., 2007; Ulrich-Lai et al., 2015). Psychiatric disorders in contemporary society frequently manifest as physiological responses and modified behavioural patterns, such as anxiety-like behaviours, which are accompanied by sensations of unease and apprehension (Ma, 2013).

5.9 SLEEP DEPRIVATION AND DEPRESSION

Sleep is essential for regulating body temperature, metabolism, maintaining internal balance, regulating hormones, and ensuring the proper functioning of all biological processes. Insufficient or disruptive sleep is a dependable indication and a factor that increases the likelihood of (serious) depression (Harvey et al., 2011; Roberts and Duong, 2014; Ramirez-Mahaluf et al., 2020; Kurczewska et al., 2021). Sleep disruptions refer to a broad spectrum of issues, such as trouble initiating or maintaining sleep and low sleep quality. Sleep disruption is a strong predictor of developing depression in the general population (Baglioni et al., 2011; Lee et al., 2013; Drake et al., 2014; Conklin et al., 2018). A minimum of 65% of patients experiencing serious depression report experiencing sleep disturbances, such as getting up early in the morning and having trouble falling asleep and/or staying asleep (Asarnow and Mirchandaney, 2024). Sleep disturbance is a significant aspect of the diagnostic criteria for major depression, as it is a strong and consistent occurrence. In serious depression, the structure of sleep stages throughout the night is changed, not just causing sleep disturbances. This encompasses a more rapid initiation of the initial REM sleep episode, along with a general augmentation in the duration of REM sleep (Freeman et al., 2020). Furthermore, the inhibition of REM sleep is a prevalent characteristic of various antidepressant medications, indicating that alterations in REM sleep can contribute to the antidepressant impact of these treatments. Antidepressant medicines can normalize the duration and onset of REM sleep in depressed patients, resulting in quicker entry into REM sleep and increased volume of REM sleep. Alternatively, anomalies in REM sleep may not directly cause depression but instead suggest deficiencies in the regulation of NREM sleep. In other words, a decrease in NREM slow-wave sleep might increase REM sleep, leading to a quicker transition into REM sleep during the early part of the night (Hemmeter et al., 2010). Consistent with this hypothesis, patients suffering from serious depression exhibit reduced NREM, slow-wave sleep, and the corresponding slow-wave activity (SWA) (Freeman et al., 2020). Randomized controlled trials have shown that SD can provide an antidepressant effect in around 45% of patients (Wu et al., 2001). Regrettably, this advantage is temporary as over 80% of these individuals experience a relapse into depression after recovering from sleep (Hemmeter et al., 2010). One potential explanation for the positive effect of this antidepressant on sleep loss is the restoration of normal connections in the default mode network (Bosch et al., 2013). Depression is increasingly prevalent worldwide, and current epidemiological data indicate that it is currently the third most significant factor in the global disease burden (Gustavsson et al., 2011). Approximately 25% of individuals in developed nations will experience depression at least once in their lifetime (Collins et al., 2011). According to reports, a majority of medical students in the United States, specifically 51% 59%, experienced poor sleep quality (Brick et al., 2010) and Lithuania (Preisegolaviciute et al., 2010). Several studies conducted in the United States, Australia, India, and other countries have consistently demonstrated a strong correlation between inadequate sleep quality and academic underperformance, as well as higher levels of depression among students. The incidence of substandard sleep quality in this study surpasses the figures published in existing literature, with rates ranging from 30% to 59% (Petit et al., 2004; Brick et al., 2010; Preisegolaviciute et al., 2010; Born and Feld, 2012).

5.10 SLEEP DEPRIVATION AND ANXIETY

Individuals with anxiety disorders frequently report experiencing disrupted sleep as a typical subjective complaint (Carruth, 2000; Maclean and Datta, 2007). Among individuals with mental conditions, particularly anxiety disorders, insomnia is the prevailing sleep disruption that is frequently observed (Szelenberger and Soldatos, 2005).

SD induces both specific and general symptoms of anxiety in individuals who are in good health (Sagaspe et al., 2006; Babson et al., 2010), and a staggering 70% of those afflicted with anxiety problems also experience SD (Wyatt et al., 1971; Foley et al., 2004). The International Classification of Sleep Disorders (ICSD) has established a diagnostic classification system for sleep disorders. This system, particularly the third edition released in 2014 by the American Academy of Sleep Medicine, is a valuable resource for medical practitioners specializing in sleep disorders. It's important to remember that sleeplessness affects approximately 33% of adults in the United States (Ancoli-Israel and Roth, 1999). In Europe, the occurrence rate of severe insomnia ranges from 4% to 22% (Chevalier et al., 1999; Hajak and SINE Study Group. Study of Insomnia in Europe, 2001). The data from the Morfeo 1 Study revealed a prevalence of insomnia of 20% among Italian participants. Additionally, sleeplessness affected 44% of the individuals (Terzano et al., 2004). Many studies indicate that animal enter in paradoxical sleep deprivation (PSD), sometimes referred to as rapid eye movement sleep deprivation (REMSD) (Vollert et al., 2011; Pires et al., 2016). Anxiety behaviour might be caused by the total SD (Xu et al., 2010). Clinical studies indicate that there is a rise in anxiety as a result of SD. The outcomes of animal tests have displayed a lack of consistency. Dement was the first to identify angiogenesis generated by SD in humans. He documented a complicated behavioral profile resulting from REMSD. This profile consisted of a slight elevation in acute anxiety, accompanied by heightened fierceness or impulsivity and a lack of focus. The strong correlation between sleep disruption and anxiety necessitates a thorough investigation to discover potentially efficacious treatments. Hence, it is crucial to acknowledge that SD is not a stand-alone problem but rather frequently related to different conditions, such as hypertension, anxiety disorders, or depression, either as a cause or as a consequential event. This complicates the process of identifying symptoms or issues that are exclusively related to SD. Furthermore, the underlying biological mechanism responsible for these anxious behaviors produced by SD remains uncertain. This chapter also intended to assess the fundamental biological process that is responsive to changes in sleep and how it affects behavior.

5.11 SLEEP DEPRIVATION AND MEMORY

The hippocampus is a memory-storing limbic area on both sides of the brain that lies below the cerebral cortex. There is a lot of evidence from studies on both humans and animals that the hippocampus is very important for learning and memory. The most convincing and important proof came from the well-known case of a patient named R.B., who was diagnosed with profound anterograde amnesia linked to ischemia damage from open-heart surgery. Postmortem analysis of R.B.'s brain revealed the presence of bilateral lesions specifically located in the cornu ammonis region 1 (CA1) of the hippocampus (Zola-Morgan et al., 1986; Franklin and Grossberg, 2017; Harrington et al., 2018). These findings unequivocally demonstrate that the impairment of the CA1 region is sufficient to induce memory impairments. Moreover, the animals experienced significant impairments in spatial memory in both the water maze and radial arm maze as a result of the deficit of the hippocampus. Depression models additionally demonstrate memory impairments, and a reduction in the size of the hippocampus is frequently reported in individuals with serious depression. The hippocampus plays a crucial role in several cognitive processes, such as learning and memory formation, memory encoding and retrieval, regulation of emotional behaviours, and anxiety by receiving sensory input from different cortical regions. Mounting data from several levels of analysis, including phylogenetic and descriptive, indicate that sleep is crucial for learning and memory consolidation, as well as for maintaining adequate hippocampus function. Several studies have utilized SD analysis to investigate the impact of sleep on hippocampus function. SD may impact many stages of memory consolidation. Previous SD can have a detrimental impact on the process of memory formation, while SD following learning appears to disrupt the process of memory consolidation. Multiple research conducted on both humans and animals has consistently demonstrated that SD before learning negatively affects the process of memory acquisition and encoding. This indicates that SD has detrimental effects on learning and memory that rely on the hippocampus. After acquiring new information, SD also hampers the process of consolidating memories. Therefore, the hippocampus seems to be vulnerable to this injury, as evidenced by the impact on memory tasks that rely on the hippocampus after SD. In previous studies, mice who were deprived of sleep had difficulties in memory function when tested in a memory challenge that relies on the hippocampus, known as the Morris water maze (MWM). Mice that experienced SD following training exhibit cognitive deficits in the MWM (Saygin et al., 2017). Previous research has documented the presence of SD before the acquisition of fear responses in conditioning experiments, with consistent findings across different studies (Qureshi and Jha, 2017). Each stage of memory formation necessitates a precise sequence of molecular changes, and SD can interfere with this chemical process. Memory is most susceptible to the negative effects of SD when sleep is sacrificed thereafter procurement. Similarly, research conducted on people also supports the idea that sleep prior to learning plays a crucial role in priming the brain for memory formation, and SD before learning hampers the process of memory formation. Hence, sleep is equally crucial not just for the process of storing and strengthening memories, but also for priming the brain for the creation of new memories the following day.

5.12 SLEEP DEPRIVATION AND MEMORY CONSOLIDATION

An intriguing hypothesis suggests that sleep plays a key role in the process of memory development. Multiple research has examined the concept that sleep following learning enhances memory consolidation. Similarly, multiple research indicates that sleep plays a role in the process of strengthening neural connections after the development of plasticity during awake (Peigneux et al., 2001). Considering the importance of sleep in memory function, multiple studies indicate that sleep loss has a detrimental impact on cognitive abilities, such as attention, learning, and both working and reference memory. Consistent with this, research using functional magnetic resonance imaging (fMRI) in humans demonstrates that sleep specifically enhances hippocampus activity following the acquisition of new information (Peigneux et al., 2001). Sleep plays a key role in enhancing hippocampal-dependent memory formation. Multiple research supports the notion that sleep aids in the process of memory consolidation, whereas inadequate sleep has detrimental effects on this particular phase of memory. Moreover, there is substantial discussion regarding the association between disrupted sleep, cognitive impairments, and depressive symptoms. It is generally observed that models of depression often exhibit shrinkage in the hippocampus (Campbell and Macqueen, 2004). Moreover, SD appears to expedite the emergence of depressive symptoms and cognitive deficits.

5.13 PREVALENCE OF SLEEP DEPRIVATION

SD is an escalating issue among adolescents. Adolescent SD has been referred to as a significant global issue of great magnitude (Foley et al., 2004; Marshall and Lallukka, 2018). SD is linked to a heightened likelihood of experiencing despair, anxiety, suicidal thoughts, and self-inflicted damage (Gangwisch et al., 2010; Clarke and Harvey, 2012; Hysing et al., 2015). In addition, sleep-deprived adolescents demonstrate diminished focus, impaired cognitive abilities, and signs of excessive daytime sleepiness (EDS) (Gangwisch et al., 2010; Lo et al., 2016). Adolescents have been consistently reporting a reduction in the amount of time they sleep. The prevalence of students consistently obtaining less than seven hours of sleep during the night decreased from 72% in 1991 to 63% in 2012 in the United States (Bierman, 2021). Over the past 50 years, the amount of time that American adults and teenagers spend sleeping has reduced by 1.5–2 hours (Van Cauter et al., 2008). More than 30% of Americans aged 30–64 reported sleeping for less than six hours, indicating a subjective sleep length. According to the National Health Institute Survey (NHIS) conducted between 2004 and 2006, over 70 million Americans are affected by chronic sleep disruptions and sleep disorders (Foley et al., 2004). In Europe, the HELENA study found that 33% of teenagers in Europe have insufficient sleep (Rey-Lopez et al., 2014). According to a recent poll done on the Chinese population, 51.0% of adolescents fail to attain the recommended amount of sleep (<8.0 hour per day) on weekdays, while 9.8% fall short on weekends (Chen et al., 2019). An extensive epidemiological study conducted on Chinese high school students found that 18.0% of them experienced inadequate sleep (Tang et al., 2016; Wadhwa et al., 2017). However, the decline in sleep quality worsens as individuals age (Wang et al., 2020). Adolescents may be more affected by sleeping less than seven hours compared to adults, as they have a higher biological sleep requirement (i.e. they need more hours of sleep each day) than adults. The 2015 sleep time recommendations by the National Sleep Foundation suggest that adults aged 18–64 may find it suitable to sleep for six hours (Hirshkowitz et al., 2015). However, adolescents shouldn't have the same duration of sleep.

5.14 MINOCYCLINE'S ROLE IN INSOMNIA, NEUROINFLAMMATION, DEPRESSION, ANXIETY, AND MEMORY

In the United State, more than 25% of people have trouble sleeping, and about one-third of adults report having at least one episode of SD (Ohayon, 2002). Furthermore, in psychiatric populations, sleep loss is among the very common complaints, encompassing adults diagnosed with major depressive disorder (Irwin et al., 2015). It is mentioned in 50%–70% of individuals receiving primary care (Tubtimtes et al., 2009). Over the past decade, a growing body of research has displayed that sleeplessness is not just a medical condition, but also a risk factor for inflammatory and infectious disorders, as well as depression, and all-cause mortality (Irwin, 2015). Epidemiological data indicate that sleep disturbances (such as wakefulness, poor sleep quality, and insufficient sleep) raise the risk of inflammatory diseases with both inflammation and sleep disturbances potentially contributing to depression. This highlights the connection between sleep and immune system (Gorman, 1996; Irwin, 2015). The homeostatic function of sleep in the control of inflammatory biological dynamics is also covered, as is the SD effect on the innate immune and molecular level of the system. Additionally, there is a mutual relationship between sleep and the innate immune system. Translational animal models have provided strong evidence that inflammatory cytokines homeostatically regulate sleep. Inflammation and inflammatory signaling are triggered by sleep loss, and self-reported sleep disturbance (i.e., complaints of insomnia) is linked to elevated inflammation. It is believed that these heightened inflammatory levels lead to an increase in inflammatory markers (Mouat et al., 2022). Complaints of insomnia cause inflammation (Irwin, 2015). Second-generation tetracycline minocycline (7-dimethylamino-6-dimethyl-6-deoxytetracycline) is efficient against both Gram-positive and Gram-negative bacteria, particularly *Staphylococcus aureus* (Garrido-Mesa et al., 2013). The U.S. Food and Drug Administration (FDA) has approved the tetracycline antibiotic minocycline for use in treating a range of bacterial illnesses (Lodise et al., 2021). Because of its anti-inflammatory qualities, minocycline has been used to treat inflammatory conditions (Langevitz et al., 2000). There are numerous clinical studies investigating the effectiveness of using minocycline off-label to treat MDD (Rosenblat and Mcintyre, 2018). By inhibiting microglial activation, minocycline, a second-generation tetracycline, had anti-inflammatory and neuroprotective benefits. Minocycline showed effectiveness and appeared to be a promising neuroprotective medication (Malhotra, 2018). Additionally, minocycline reduced symptoms of depression in individuals with major depressive disorder (Dean et al., 2014). The symptoms of anxiety depend on stress (Zhang et al., 2019). When cerebral ischemia/reperfusion causes memory impairment in rats, minocycline has neuroprotective effects that seem to be mediated by its anti-inflammatory and antioxidative qualities (Chen et al., 2012).

5.15 DISCUSSION

Depression is a complicated illness that is most likely caused by many interrelated variables, including stress, environment, developmental impacts, lifestyle, and genetics. An increasing body of research suggests that SD may raise the risk of depression (Roberts and Duong, 2014; Meerlo et al., 2015; Li et al., 2017). Anxiety and depression progress more quickly when there is inflammation in the brain, and inflammation in the brain can be brought on by SD. Strong epidemiological research suggests that getting too little sleep or poor quality sleep increases the risk of inflammatory diseases; inflammation and SD are thought to be related to depression (Mullington et al., 2010; Irwin, 2019). A growing body of research indicates that microglia have a role in the neuroinflammation associated with depression (Deng et al., 2020; Jia et al., 2021). There is a significant density of microglial cells in the hippocampus, and studies have demonstrated that stress can activate the hippocampal microglia. Additionally, there are signs that the pathophysiology of depression is associated with hippocampal microglial activation (Gong et al., 2018; Liu et al., 2019). Crucially, one aspect of minocycline's mode of action involves the inhibition of microglial activation (Tikka et al., 2001; Kobayashi et al., 2013b). Mice with depressive symptoms and anxiety had reduced levels of Keapl and Nrf2 protein expression in the hippocampus and prefrontal cortex, respectively (Yao et al., 2016). Minocycline has the ability to upregulate Nrf2 expression in a diabetic neuropathy mice model (Shahzad et al., 2016) as well as a cellular damage model in rats (Tian et al., 2017). In individuals with major depressive illness, minocycline can also lessen depressed symptoms (Dean et al., 2014) and in animal models (Zhang et al., 2019; Wang et al., 2020). Numerous investigations have shown a strong comorbidity between anxiety and depressive phenotypes (Cryan and Holmes, 2005), and disturbed sleep has been connected to the emergence and advancement of anxiety disorders (Neckelmann et al., 2007).

5.16 FUTURE PERSPECTIVES

Our modern society is characterized by disturbed sleep. SD has significantly impacted social, economic, and health outcomes, causing a community health epidemic. Over an extended period of time, sleep loss accumulates and goes untreated, leading to a sleep-deprived and depressed society. On the other hand, persistent SD over an extended period of time can interfere with signaling pathways linked to learning and memory, inhibit neurogenesis, and potentially cause depression and physical changes to the brain, such as shrinkage of the hippocampus. Modern society is known for its lack of sleep, which can lead to serious health problems. Therefore, treatments that increase sleep may help cognitive function in the short and long term, including mood control, learning and memory, and regular brain function. Minocycline plays an essential role in neuroprotection against the harmful biochemical and behavioral effects of SD. In conclusion, treatment with minocycline shows promise in reversing changes brought on by SD; therefore, more investigation is necessary to confirm minocycline's effects on mood disorders linked to inadequate sleep in people.

REFERENCES

Abelaira, H. M., Reus, G. Z. & Quevedo, J. 2013. Animal models as tools to study the pathophysiology of depression. *Braz J Psychiatry*, 35 Suppl 2, S112–20.

Alkadhi, K., Zagaar, M., Alhaider, I., Salim, S. & Aleisa, A. 2013. Neurobiological consequences of sleep deprivation. *Curr Neuropharmacol*, 11, 231–49.

Allen, N. J. & Lyons, D. A. 2018. Glia as architects of central nervous system formation and function. *Science*, 362, 181–85.

Alonso, J., Lepine, J. P. & ESEMeD/MHEDEA 2000 Scientific Committee. 2007. Overview of key data from the European Study of the Epidemiology of Mental Disorders (ESEMeD). *J Clin Psychiatry*, 68 Suppl 2, 3–9.

Asarnow, L. D. & Mirchandaney, R. 2024. Sleep and mood disorders among youth. *Psychiatr Clin North Am*, 47, 255–72.

Atrooz, F. & Salim, S. 2020. Sleep deprivation, oxidative stress and inflammation. *Adv Protein Chem Struct Biol*, 119, 309–36.

Ancoli-Israel, S. & Roth, T. 1999. Characteristics of insomnia in the United States: Results of the 1991 National Sleep Foundation Survey. I. *Sleep*, 22(Suppl 2), S347–53.

Babson, K. A., Trainor, C. D., Feldner, M. T. & Blumenthal, H. 2010. A test of the effects of acute sleep deprivation on general and specific self-reported anxiety and depressive symptoms: An experimental extension. *J Behav Ther Exp Psychiatry*, 41, 297–303.

Baglioni, C., Battagliese, G., Feige, B., Spiegelhalder, K., Nissen, C., Voderholzer, U., Lombardo, C. & Riemann, D. 2011. Insomnia as a predictor of depression: A meta-analytic evaluation of longitudinal epidemiological studies. *J Affect Disord*, 135, 10–19.

Bellesi, M., De Vivo, L., Chini, M., Gilli, F., Tononi, G. & Cirelli, C. 2017. Sleep loss promotes astrocytic phagocytosis and microglial activation in mouse cerebral cortex. *J Neurosci*, 37, 5263–73.

Beumer, W., Gibney, S. M., Drexhage, R. C., Pont-Lezica, L., Doorduin, J., Klein, H. C., Steiner, J., Connor, T. J., Harkin, A., Versnel, M. A. & Drexhage, H. A. 2012. The immune theory of psychiatric diseases: A key role for activated microglia and circulating monocytes. *J Leukoc Biol*, 92, 959–75.

Bierman, A. 2021. Why have sleep problems in later-midlife grown following the great recession? A comparative cohort analysis. *J Gerontol B Psychol Sci Soc Sci*, 76, 1005–14.

Born, J. & Feld, G. B. 2012. Sleep to upscale, sleep to downscale: Balancing homeostasis and plasticity. *Neuron*, 75, 933–5.

Bosch, O. G., Rihm, J. S., Scheidegger, M., Landolt, H. P., Stampfli, P., Brakowski, J., Esposito, F., Rasch, B. & Seifritz, E. 2013. Sleep deprivation increases dorsal nexus connectivity to the dorsolateral prefrontal cortex in humans. *Proc Natl Acad Sci USA*, 110, 19597–602.

Brick, C. A., Seely, D. L. & Palermo, T. M. 2010. Association between sleep hygiene and sleep quality in medical students. *Behav Sleep Med*, 8, 113–21.

Buysse, D. J., Angst, J., Gamma, A., Ajdacic, V., Eich, D. & Rossler, W. 2008. Prevalence, course, and comorbidity of insomnia and depression in young adults. *Sleep*, 31, 473–80.

Campbell, S. & Macqueen, G. 2004. The role of the hippocampus in the pathophysiology of major depression. *J Psychiatry Neurosci*, 29, 417–26.

Carruth, S. G., Jr. 2000. Separation anxiety disorder: Planning treatment. *Pediatr Rev*, 21, 248.

Chapman, D. P., Presley-Cantrell, L. R., Liu, Y., Perry, G. S., Wheaton, A. G. & Croft, J. B. 2013. Frequent insufficient sleep and anxiety and depressive disorders among U.S. Community dwellers in 20 States, 2010. *Psychiatr Serv*, 64, 385–7.

Chen, Y., Li, M., Zheng, Y. & Yang, L. 2019. Evaluation of hemisphere lateralization with bilateral local field potential recording in secondary motor cortex of mice. *J Vis Exp*, 149, e59310.

Chen, S. D., Yin, J. H., Hwang, C. S., Tang, C. M. & Yang, D. I. 2012. Anti-apoptotic and anti-oxidative mechanisms of minocycline against sphingomyelinase/ceramide neurotoxicity: Implication in Alzheimer's disease and cerebral ischemia. *Free Radic Res*, 46, 940–50.

Chevalier, H., Los, F., Boichut, D., Bianchi, M., Nutt, D. J., Hajak, G., Hetta, J., Hoffmann, G. & Crowe, C. 1999. Evaluation of severe insomnia in the general population: Results of a European multinational survey. *J Psychopharmacol*, 13, S21–24.

Choudhury, M. E., Miyanishi, K., Takeda, H., Islam, A., Matsuoka, N., Kubo, M., Matsumoto, S., Kunieda, T., Nomoto, M., Yano, H. & Tanaka, J. 2020. Phagocytic elimination of synapses by microglia during sleep. *Glia*, 68, 44–59.

Clarke, G. & Harvey, A. G. 2012. The complex role of sleep in adolescent depression. *Child Adolesc Psychiatr Clin N Am*, 21, 385–400.

Collins, P. Y., Patel, V., Joestl, S. S., March, D., Insel, T. R., Daar, A. S., Scientific Advisory, B., Anderson, W., Dhansay, M. A., Phillips, A., Shurin, S., Walport, M., Ewart, W., Savill, S. J., Bordin, I. A., Costello, E. J., Durkin, M., Fairburn, C., Glass, R. I., Hall, W., Huang, Y., Hyman, S. E., Jamison, K., Kaaya, S., Kapur, S., Kleinman, A., Ogunniyi, A., Otero-Ojeda, A., Poo, M. M., Ravindranath, V., Sahakian, B. J., Saxena, S., Singer, P. A. Stein, D. J. et al. 2011. Grand challenges in global mental health. *Nature*, 475, 27–30.

Conklin, A. I., Yao, C. A. & Richardson, C. G. 2018. Chronic sleep deprivation and gender-specific risk of depression in adolescents: A prospective population-based study. *BMC Public Health*, 18, 724.

Cryan, J. F. & Holmes, A. 2005. The ascent of mouse: Advances in modelling human depression and anxiety. *Nat Rev Drug Discov*, 4, 775–90.

D'mello, C., Le, T. & Swain, M. G. 2009. Cerebral microglia recruit monocytes into the brain in response to tumor necrosis factoralpha signaling during peripheral organ inflammation. *J Neurosci*, 29, 2089–102.

Dantzer, R., O'connor, J. C., Freund, G. G., Johnson, R. W. & Kelley, K. W. 2008. From inflammation to sickness and depression: When the immune system subjugates the brain. *Nat Rev Neurosci*, 9, 46–56.

Dean, O. M., Maes, M., Ashton, M., Berk, L., Kanchanatawan, B., Sughondhabirom, A., Tangwongchai, S., Ng, C., Dowling, N., Malhi, G. S. & Berk, M. 2014. Protocol and rationale-the efficacy of minocycline as an adjunctive treatment for major depressive disorder: A double blind, randomised, placebo controlled trial. *Clin Psychopharmacol Neurosci*, 12, 180–8.

Deng, S. L., Chen, J. G. & Wang, F. 2020. Microglia: A central player in depression. *Curr Med Sci*, 40, 391–400.

Ding, X., Yue, X., Zheng, R., Bi, C., Li, D. & Yao, G. 2019. Classifying major depression patients and healthy controls using EEG, eye tracking and galvanic skin response data. *J Affect Disord*, 251, 156–61.

Drake, C. L., Pillai, V. & Roth, T. 2014. Stress and sleep reactivity: A prospective investigation of the stress-diathesis model of insomnia. *Sleep*, 37, 1295–304.

Dworak, M., Kim, T., Mccarley, R. W. & Basheer, R. 2011. Sleep, brain energy levels, and food intake: Relationship between hypothalamic ATP concentrations, food intake, and body weight during sleep-wake and sleep deprivation in rats. *Somnologie (Berl)*, 15, 111–7.

Farooqui, Z., Mohammad, R. S., Lokhandwala, M. F. & Banday, A. A. 2021. Nrf2 inhibition induces oxidative stress, renal inflammation and hypertension in mice. *Clin Exp Hypertens*, 43, 175–80.

Ferrero-Miliani, L., Nielsen, O. H., Andersen, P. S. & Girardin, S. E. 2007. Chronic inflammation: Importance of NOD2 and NALP3 in interleukin-1beta generation. *Clin Exp Immunol*, 147, 227–35.

Foley, D., Ancoli-Israel, S., Britz, P. & Walsh, J. 2004. Sleep disturbances and chronic disease in older adults: Results of the 2003 National Sleep Foundation Sleep in America Survey. *J Psychosom Res*, 56, 497–502.

Franklin, D. J. & Grossberg, S. 2017. A neural model of normal and abnormal learning and memory consolidation: Adaptively timed conditioning, hippocampus, amnesia, neurotrophins, and consciousness. *Cogn Affect Behav Neurosci*, 17, 24–76.

Freeman, D., Sheaves, B., Waite, F., Harvey, A. G. & Harrison, P. J. 2020. Sleep disturbance and psychiatric disorders. *Lancet Psychiatry*, 7, 628–37.

Frick, L. R., Williams, K. & Pittenger, C. 2013. Microglial dysregulation in psychiatric disease. *Clin Dev Immunol*, 2013, 608654.

Gangwisch, J. E., Babiss, L. A., Malaspina, D., Turner, J. B., Zammit, G. K. & Posner, K. 2010. Earlier parental set bedtimes as a protective factor against depression and suicidal ideation. *Sleep*, 33, 97–106.

Gao, T., Wang, Z., Dong, Y., Cao, J., Lin, R., Wang, X., Yu, Z. & Chen, Y. 2019. Role of melatonin in sleep deprivation-induced intestinal barrier dysfunction in mice. *J Pineal Res*, 67, e12574.

Garrido-Mesa, N., Zarzuelo, A. & Galvez, J. 2013. Minocycline: Far beyond an antibiotic. *Br J Pharmacol*, 169, 337–52.

Ghosh, A. K., O'brien, M., Mau, T. & Yung, R. 2017. Toll-like receptor 4 (TLR4) deficient mice are protected from adipose tissue inflammation in aging. *Aging (Albany NY)*, 9, 1971–82.

Gong, Y., Tong, L., Yang, R., Hu, W., Xu, X., Wang, W., Wang, P., Lu, X., Gao, M., Wu, Y., Xu, X., Zhang, Y., Chen, Z. & Huang, C. 2018. Dynamic changes in hippocampal microglia contribute to depressive-like behavior induced by early social isolation. *Neuropharmacology*, 135, 223–33.

Gore, P. R., Prajapati, C. P., Mahajan, U. B., Goyal, S. N., Belemkar, S., Ojha, S. & Patil, C. R. 2016. Protective effect of thymoquinone against cyclophosphamide-induced hemorrhagic cystitis through inhibiting DNA damage and upregulation of Nrf2 expression. *Int J Biol Sci*, 12, 944–53.

Gorman, J. M. 1996. Comorbid depression and anxiety spectrum disorders. *Depress Anxiety*, 4, 160–8.

Gustavsson, A., Svensson, M., Jacobi, F., Allgulander, C., Alonso, J., Beghi, E., Dodel, R., Ekman, M., Faravelli, C., Fratiglioni, L., Gannon, B., Jones, D. H., Jennum, P., Jordanova, A., Jonsson, L., Karampampa, K., Knapp, M., Kobelt, G., Kurth, T., Lieb, R., Linde, M., Ljungcrantz, C., Maercker, A., Melin, B., Moscarelli, M., Musayev, A., Norwood, F., Preisig, M., Pugliatti, M., Rehm, J., Salvador-Carulla, L., Schlehofer, B., Simon, R., Steinhausen, H. C., Stovner, L. J., Vallat, J. M., Van Den Bergh, P., Van Os, J., Vos, P., Xu, W., Wittchen, H. U., Jonsson, B., Olesen, J. & Group, C. D. 2011. Cost of disorders of the brain in Europe 2010. *Eur Neuropsychopharmacol*, 21, 718–79.

Hafner, M., Stepanek, M., Taylor, J., Troxel, W. M. & Van Stolk, C. 2017. Why sleep matters-the economic costs of insufficient sleep: A cross-country comparative analysis. *Rand Health Q*, 6, 11.

Hajak, G. & SINE Study Group. Study of Insomnia in Europe. 2001. Epidemiology of severe insomnia and its consequences in Germany. *Eur Arch Psychiatry Clin Neurosci*, 251, 49–56.

Halaris, A. 2019. Inflammation and depression but where does the inflammation come from? *Curr Opin Psychiatry*, 32, 422–28.

Han, Y. Y., Jin, K., Pan, Q. S., Li, B., Wu, Z. Q., Gan, L., Yang, L. & Long, C. 2020. Microglial activation in the dorsal striatum participates in anxiety-like behavior in Cyld knockout mice. *Brain Behav Immun*, 89, 326–38.

Harrington, M. O., Nedberge, K. M. & Durrant, S. J. 2018. The effect of sleep deprivation on emotional memory consolidation in participants reporting depressive symptoms. *Neurobiol Learn Mem*, 152, 10–19.

Harvey, A. G., Murray, G., Chandler, R. A. & Soehner, A. 2011. Sleep disturbance as transdiagnostic: Consideration of neurobiological mechanisms. *Clin Psychol Rev*, 31, 225–35.

Hashimoto, K. 2009. Emerging role of glutamate in the pathophysiology of major depressive disorder. *Brain Res Rev*, 61, 105–23.

Hashimoto, K. 2015. Inflammatory biomarkers as differential predictors of antidepressant response. *Int J Mol Sci*, 16, 7796–801.

Hashimoto, K. 2018. Essential role of Keap1-Nrf2 signaling in mood disorders: Overview and future perspective. *Front Pharmacol*, 9, 1182.

Hemmeter, U. M., Hemmeter-Spernal, J. & Krieg, J. C. 2010. Sleep deprivation in depression. *Expert Rev Neurother*, 10, 1101–15.

Hong, H., Kim, B. S. & Im, H. I. 2016. Pathophysiological role of neuroinflammation in neurodegenerative diseases and psychiatric disorders. *Int Neurourol J*, 20, S2–S7.

Huyett, P., Siegel, N. & Bhattacharyya, N. 2021. Prevalence of sleep disorders and association with mortality: Results from the NHANES 2009–2010. *Laryngoscope*, 131, 686–9.

Hysing, M., Sivertsen, B., Stormark, K. M. & O'connor, R. C. 2015. Sleep problems and self-harm in adolescence. *Br J Psychiatry*, 207, 306–12.

Hirshkowitz, M., Whiton, K., Albert, S. M., Alessi, C., Bruni, O., Doncarlos, L., Hazen, N., Herman, J., Adams Hillard, P. J., Katz, E. S., Kheirandish-Gozal, L., Neubauer, D. N., O'donnell, A. E., Ohayon, M., Peever, J., Rawding, R., Sachdeva, R. C., Setters, B., Vitiello, M. V. & Ware, J. C. 2015. National sleep foundation's updated sleep duration recommendations: Final report. Sleep Health, 1, 233–43.

Iliff, J. J., Wang, M., Liao, Y., Plogg, B. A., Peng, W., Gundersen, G. A., Benveniste, H., Vates, G. E., Deane, R., Goldman, S. A., Nagelhus, E. A. & Nedergaard, M. 2012. A paravascular pathway facilitates CSF flow through the brain parenchyma and the clearance of interstitial solutes, including amyloid beta. *Sci Transl Med*, 4, 147ra111.

Irwin, M. R. 2015. Why sleep is important for health: A psychoneuroimmunology perspective. *Annu Rev Psychol*, 66, 143–72.

Irwin, M. R. 2019. Sleep and inflammation: Partners in sickness and in health. *Nat Rev Immunol*, 19, 702–15.

Irwin, M. R., Olmstead, R., Breen, E. C., Witarama, T., Carrillo, C., Sadeghi, N., Arevalo, J. M., Ma, J., Nicassio, P., Bootzin, R. & Cole, S. 2015. Cognitive behavioral therapy and tai chi reverse cellular and genomic markers of inflammation in late-life insomnia: A randomized controlled trial. *Biol Psychiatry*, 78, 721–9.

Javaheripour, N., Shahdipour, N., Noori, K., Zarei, M., Camilleri, J. A., Laird, A. R., Fox, P. T., Eickhoff, S. B., Eickhoff, C. R., Rosenzweig, I., Khazaie, H. & Tahmasian, M. 2019. Functional brain alterations in acute sleep deprivation: An activation likelihood estimation meta-analysis. *Sleep Med Rev*, 46, 64–73.

Jia, X., Gao, Z. & Hu, H. 2021. Microglia in depression: Current perspectives. *Sci China Life Sci*, 64, 911–25.

Kang, A., Hao, H., Zheng, X., Liang, Y., Xie, Y., Xie, T., Dai, C., Zhao, Q., Wu, X., Xie, L. & Wang, G. 2011. Peripheral anti-inflammatory effects explain the ginsenosides paradox between poor brain distribution and anti-depression efficacy. *J Neuroinflammation*, 8, 100.

Kang, S. S., Kurti, A., Fair, D. A. & Fryer, J. D. 2014. Dietary intervention rescues maternal obesity induced behavior deficits and neuroinflammation in offspring. *J Neuroinflammation*, 11, 156.

Katoh, Y., Itoh, K., Yoshida, E., Miyagishi, M., Fukamizu, A. & Yamamoto, M. 2001. Two domains of Nrf2 cooperatively bind CBP, a CREB binding protein, and synergistically activate transcription. *Genes Cells*, 6, 857–68.

Kettenmann, H., Hanisch, U. K., Noda, M. & Verkhratsky, A. 2011. Physiology of microglia. *Physiol Rev*, 91, 461–553.

Kettenmann, H., Kirchhoff, F. & Verkhratsky, A. 2013. Microglia: New roles for the synaptic stripper. *Neuron*, 77, 10–18.

Kobayashi, K., Imagama, S., Ohgomori, T., Hirano, K., Uchimura, K., Sakamoto, K., Hirakawa, A., Takeuchi, H., Suzumura, A., Ishiguro, N. & Kadomatsu, K. 2013b. Minocycline selectively inhibits M1 polarization of microglia. *Cell Death Dis*, 4, e525.

Kobayashi, E., Suzuki, T. & Yamamoto, M. 2013a. Roles Nrf2 plays in myeloid cells and related disorders. *Oxid Med Cell Longev*, 2013, 529219.

Konsman, J. P. 2019. Inflammation and depression: A nervous plea for psychiatry to not become immune to interpretation. *Pharmaceuticals (Basel)*, 12(1), 29.

Kurczewska, E., Ferensztajn-Rochowiak, E., Rybakowski, F., Michalak, M. & Rybakowski, J. 2021. Treatment-resistant depression: Neurobiological correlates and the effect of sleep deprivation with sleep phase advance for the augmentation of pharmacotherapy. *World J Biol Psychiatry*, 22, 58–69.

Langevitz, P., Livneh, A., Bank, I. & Pras, M. 2000. Benefits and risks of minocycline in rheumatoid arthritis. *Drug Saf*, 22, 405–14.

Lee, E., Cho, H. J., Olmstead, R., Levin, M. J., Oxman, M. N. & Irwin, M. R. 2013. Persistent sleep disturbance: A risk factor for recurrent depression in community-dwelling older adults. *Sleep*, 36, 1685–91.

Li, Y., Wu, Y., Zhai, L., Wang, T., Sun, Y. & Zhang, D. 2017. Longitudinal association of sleep duration with depressive symptoms among middle-aged and older Chinese. *Sci Rep*, 7, 11794.

Liu, L. L., Li, J. M., Su, W. J., Wang, B. & Jiang, C. L. 2019. Sex differences in depressive-like behaviour may relate to imbalance of microglia activation in the hippocampus. *Brain Behav Immun*, 81, 188–97.

Lo, J. C., Ong, J. L., Leong, R. L., Gooley, J. J. & Chee, M. W. 2016. Cognitive performance, sleepiness, and mood in partially sleep deprived adolescents: The need for sleep study. *Sleep*, 39, 687–98.

Lodise, T. P., Van Wart, S., Sund, Z. M., Bressler, A. M., Khan, A., Makley, A. T., Hamad, Y., Salata, R. A., Silveira, F. P., Sims, M. D., Kabchi, B. A., Saad, M. A., Brown, C., Oler, R. E. Jr., Fowler, V. Jr. & Wunderink, R. G. 2021. Pharmacokinetic and pharmacodynamic profiling of minocycline for injection following a single infusion in critically ill adults in a phase IV open-label multicenter study (ACUMIN). *Antimicrob Agents Chemother*, 65.

Ma, Q. 2013. Role of Nrf2 in oxidative stress and toxicity. *Annu Rev Pharmacol Toxicol*, 53, 401–26.

Mackiewicz, M., Shockley, K. R., Romer, M. A., Galante, R. J., Zimmerman, J. E., Naidoo, N., Baldwin, D. A., Jensen, S. T., Churchill, G. A. & Pack, A. I. 2007. Macromolecule biosynthesis: A key function of sleep. *Physiol Genomics*, 31, 441–57.

Maclean, R. R. & Datta, S. 2007. The relationship between anxiety and sleep-wake behavior after stressor exposure in the rat. *Brain Res*, 1164, 72–80.

Maes, M., Yirmyia, R., Noraberg, J., Brene, S., Hibbeln, J., Perini, G., Kubera, M., Bob, P., Lerer, B. & Maj, M. 2009. The inflammatory & neurodegenerative (I&ND) hypothesis of depression: Leads for future research and new drug developments in depression. *Metab Brain Dis*, 24, 27–53.

Mahato, S. & Paul, S. 2019. Classification of depression patients and normal subjects based on electroencephalogram (EEG) signal using alpha power and theta asymmetry. *J Med Syst*, 44, 28.

Malhotra, R. K. 2018. Neurodegenerative disorders and sleep. *Sleep Med Clin*, 13, 63–70.

Marshall, N. S. & Lallukka, T. 2018. Sleep pirates—Are we really living through a sleep deprivation epidemic and what's stealing our sleep? *Eur J Public Health*, 28, 394–5.

McDonald, D. M. & Keene, A. C. 2010. The sleep-feeding conflict: Understanding behavioral integration through genetic analysis in drosophila. *Aging (Albany NY)*, 2, 519–22.

Meerlo, P., Havekes, R. & Steiger, A. 2015. Chronically restricted or disrupted sleep as a causal factor in the development of depression. *Curr Top Behav Neurosci*, 25, 459–81.

Miller, A. H., Haroon, E. & Felger, J. C. 2017. Therapeutic implications of brain-immune interactions: Treatment in translation. *Neuropsychopharmacology*, 42, 334–59.

Miller, A. H. & Raison, C. L. 2016. The role of inflammation in depression: From evolutionary imperative to modern treatment target. *Nat Rev Immunol*, 16, 22–34.

Mouat, I. C., Goldberg, E. & Horwitz, M. S. 2022. Age-associated B cells in autoimmune diseases. *Cell Mol Life Sci*, 79, 402.

Mullin, B. C., Phillips, M. L., Siegle, G. J., Buysse, D. J., Forbes, E. E. & Franzen, P. L. 2013. Sleep deprivation amplifies striatal activation to monetary reward. *Psychol Med*, 43, 2215–25.

Mullington, J. M., Simpson, N. S., Meier-Ewert, H. K. & Haack, M. 2010. Sleep loss and inflammation. *Best Pract Res Clin Endocrinol Metab*, 24, 775–84.

Neckelmann, D., Mykletun, A. & Dahl, A. A. 2007. Chronic insomnia as a risk factor for developing anxiety and depression. *Sleep*, 30, 873–80.

Ohayon, M. M. 2002. Epidemiology of insomnia: What we know and What we still need to learn. *Sleep Med Rev*, 6, 97–111.

Olah, M., Biber, K., Vinet, J. & Boddeke, H. W. 2011. Microglia phenotype diversity. *CNS Neurol Disord Drug Targets*, 10, 108–18.

Pae, C. U., Marks, D. M., Han, C. & Patkar, A. A. 2008. A case of transient hallucination with ropinirole augmentation antidepressant in a patient with treatment-resistant depression: Is there differential effect of ropinirole dose on developing psychotic symptom? *Prog Neuropsychopharmacol Biol Psychiatry*, 32, 1087–8.

Peigneux, P., Laureys, S., Delbeuck, X. & Maquet, P. 2001. Sleeping brain, learning brain. The role of sleep for memory systems. *NeuroReport*, 12, A111–24.

Petit, D., Gagnon, J. F., Fantini, M. L., Ferini-Strambi, L. & Montplaisir, J. 2004. Sleep and quantitative EEG in neurodegenerative disorders. *J Psychosom Res*, 56, 487–96.

Pires, G. N., Bezerra, A. G., Tufik, S. & Andersen, M. L. 2016. Effects of experimental sleep deprivation on anxiety-like behavior in animal research: Systematic review and meta-analysis. *Neurosci Biobehav Rev*, 68, 575–89.

Preisegolaviciute, E., Leskauskas, D. & Adomaitiene, V. 2010. Associations of quality of sleep with lifestyle factors and profile of studies among Lithuanian students. *Medicina (Kaunas)*, 46, 482–9.

Prinz, M. & Priller, J. 2014. Microglia and brain macrophages in the molecular age: From origin to neuropsychiatric disease. *Nat Rev Neurosci*, 15, 300–12.

Qureshi, M. F. & Jha, S. K. 2017. Short-term total sleep-deprivation impairs contextual fear memory, and contextual fear-conditioning reduces REM sleep in moderately anxious Swiss mice. *Front Behav Neurosci*, 11, 239.

Ramirez-Mahaluf, J. P., Rozas-Serri, E., Ivanovic-Zuvic, F., Risco, L. & Vohringer, P. A. 2020. Effectiveness of sleep deprivation in treating acute bipolar depression as augmentation strategy: A systematic review and meta-analysis. *Front Psychiatry*, 11, 70.

Rey-Lopez, J. P., De Carvalho, H. B., De Moraes, A. C., Ruiz, J. R., Sjostrom, M., Marcos, A., Polito, A., Gottrand, F., Manios, Y., Kafatos, A., Molnar, D., Widhalm, K., De Henauw, S., Moreno, L. A. & Study, H. 2014. Sleep time and cardiovascular risk factors in adolescents: The HELENA (Healthy Lifestyle in Europe by Nutrition in Adolescence) study. *Sleep Med*, 15, 104–10.

Roberts, R. E. & Duong, H. T. 2014. The prospective association between sleep deprivation and depression among adolescents. *Sleep*, 37, 239–44.

Rogers, B. G., Bainter, S. A., Smith-Alvarez, R., Wohlgemuth, W. K., Antoni, M. H., Rodriguez, A. E. & Safren, S. A. 2021. Insomnia, health, and health-related quality of life in an urban clinic sample of people living with HIV/AIDS. *Behav Sleep Med*, 19, 516–32.

Ronnback, L. & Hansson, E. 2004. On the potential role of glutamate transport in mental fatigue. *J Neuroinflammation*, 1, 22.

Rosenblat, J. D. & Mcintyre, R. S. 2018. Efficacy and tolerability of minocycline for depression: A systematic review and meta-analysis of clinical trials. *J Affect Disord*, 227, 219–25.

Sagaspe, P., Sanchez-Ortuno, M., Charles, A., Taillard, J., Valtat, C., Bioulac, B. & Philip, P. 2006. Effects of sleep deprivation on Color-Word, Emotional, and Specific Stroop interference and on self-reported anxiety. *Brain Cogn*, 60, 76–87.

Saygin, M., Ozguner, M. F., Onder, O., Doguc, D. K., Ilhan, I. & Peker, Y. 2017. The impact of sleep deprivation on hippocampal-mediated learning and memory in rats. *Bratisl Lek Listy*, 118, 408–16.

Shahzad, K., Bock, F., Al-Dabet, M. M., Gadi, I., Nazir, S., Wang, H., Kohli, S., Ranjan, S., Mertens, P. R., Nawroth, P. P. & Isermann, B. 2016. Stabilization of endogenous Nrf2 by minocycline protects against Nlrp3-inflammasome induced diabetic nephropathy. *Sci Rep*, 6, 34228.

Shi, A., Shi, H., Wang, Y., Liu, X., Cheng, Y., Li, H., Zhao, H., Wang, S. & Dong, L. 2018. Activation of Nrf2 pathway and inhibition of NLRP3 inflammasome activation contribute to the protective effect of chlorogenic acid on acute liver injury. *Int Immunopharmacol*, 54, 125–30.

Slama, H., Chylinski, D. O., Deliens, G., Leproult, R., Schmitz, R. & Peigneux, P. 2018. Sleep deprivation triggers cognitive control impairments in task-goal switching. *Sleep*, 41(2), zsx200.

Soczynska, J. K., Mansur, R. B., Brietzke, E., Swardfager, W., Kennedy, S. H., Woldeyohannes, H. O., Powell, A. M., Manierka, M. S. & Mcintyre, R. S. 2012. Novel therapeutic targets in depression: Minocycline as a candidate treatment. *Behav Brain Res*, 235, 302–17.

Spiers, J. G., Breda, C., Robinson, S., Giorgini, F. & Steinert, J. R. 2019. Drosophila Nrf2/Keap1 mediated redox signaling supports synaptic function and longevity and impacts on circadian activity. *Front Mol Neurosci*, 12, 86.

Suzuki, T., Motohashi, H. & Yamamoto, M. 2013. Toward clinical application of the Keap1-Nrf2 pathway. *Trends Pharmacol Sci*, 34, 340–6.

Suzuki, T. & Yamamoto, M. 2015. Molecular basis of the Keap1-Nrf2 system. *Free Radic Biol Med*, 88, 93–100.

Szelenberger, W. & Soldatos, C. 2005. Sleep disorders in psychiatric practice. *World Psychiatry*, 4, 186–90.

Tang, D., Li, P., Guo, L., Xu, Y., Gao, X., Deng, J., Huang, J., Huang, G., Wu, H., Yue, Y. & Lu, C. 2016. The prevalences of and association between nonmedical prescription opioid use and poor sleep among Chinese high school students. *Sci Rep*, 6, 30411.

Terzano, M. G., Parrino, L., Cirignotta, F., Ferini-Strambi, L., Gigli, G., Rudelli, G. & & Sergio Sommacal; Studio Morfeo Committee. 2004. Studio Morfeo: Insomnia in primary care, a survey conducted on the Italian population. *Sleep Med*, 5, 67–75.

Tian, Y., Wu, X., Guo, S., Ma, L., Huang, W. & Zhao, X. 2017. Minocycline attenuates sevoflurane-induced cell injury via activation of Nrf2. *Int J Mol Med*, 39, 869–78.

Tikka, T., Fiebich, B. L., Goldsteins, G., Keinanen, R. & Koistinaho, J. 2001. Minocycline, a tetracycline derivative, is neuroprotective against excitotoxicity by inhibiting activation and proliferation of microglia. *J Neurosci*, 21, 2580–8.

Tubtimtes, S., Sukying, C. & Prueksaritanond, S. 2009. Sleep problems in out-patient of primary care unit. *J Med Assoc Thai*, 92, 273–8.

Ulrich-Lai, Y. M., Fulton, S., Wilson, M., Petrovich, G. & Rinaman, L. 2015. Stress exposure, food intake and emotional state. *Stress*, 18, 381–99.

Van Cauter, E., Spiegel, K., Tasali, E. & Leproult, R. 2008. Metabolic consequences of sleep and sleep loss. *Sleep Med*, 9 Suppl 1, S23–8.

Vollert, C., Zagaar, M., Hovatta, I., Taneja, M., Vu, A., Dao, A., Levine, A., Alkadhi, K. & Salim, S. 2011. Exercise prevents sleep deprivation-associated anxiety-like behavior in rats: Potential role of oxidative stress mechanisms. *Behav Brain Res*, 224, 233–40.

Wadhwa, M., Prabhakar, A., Ray, K., Roy, K., Kumari, P., Jha, P. K., Kishore, K., Kumar, S. & Panjwani, U. 2017. Inhibiting the microglia activation improves the spatial memory and adult neurogenesis in rat hippocampus during 48 h of sleep deprivation. *J Neuroinflammation*, 14, 222.

Wang, B., Huang, X., Pan, X., Zhang, T., Hou, C., Su, W. J., Liu, L. L., Li, J. M. & Wang, Y. X. 2020. Minocycline prevents the depressive-like behavior through inhibiting the release of HMGB1 from microglia and neurons. *Brain Behav Immun*, 88, 132–43.

Wible, R. S., Ramanathan, C., Sutter, C. H., Olesen, K. M., Kensler, T. W., Liu, A. C. & Sutter, T. R. 2018. NRF2 regulates core and stabilizing circadian clock loops, coupling redox and timekeeping in Mus Musculus. *eLife*, 7, 31656.

Wisor, J. P., Schmidt, M. A. & Clegern, W. C. 2011. Evidence for neuroinflammatory and microglial changes in the cerebral response to sleep loss. *Sleep*, 34, 261–72.

Woody, C. A., Ferrari, A. J., Siskind, D. J., Whiteford, H. A. & Harris, M. G. 2017. A systematic review and meta-regression of the prevalence and incidence of perinatal depression. *J Affect Disord*, 219, 86–92.

Wohleb, E. S., Fenn, A. M., Pacenta, A. M., Powell, N. D., Sheridan, J. F. & Godbout, J. P. 2012. Peripheral innate immune challenge exaggerated microglia activation, increased the number of inflammatory CNS macrophages, and prolonged social withdrawal in socially defeated mice. *Psychoneuroendocrinology*, 37, 1491–505.

Wohleb, E. S., Patterson, J. M., Sharma, V., Quan, N., Godbout, J. P. & Sheridan, J. F. 2014. Knockdown of interleukin-1 receptor type-1 on endothelial cells attenuated stress-induced neuroinflammation and prevented anxiety-like behavior. *J Neurosci*, 34, 2583–91.

Wolf, S. A., Boddeke, H. W. & Kettenmann, H. 2017. Microglia in physiology and disease. *Annu Rev Physiol*, 79, 619–43.

Wyatt, R. J., Fram, D. H., Kupfer, D. J. & Snyder, F. 1971. Total prolonged drug-induced REM sleep suppression in anxious-depressed patients. *Arch Gen Psychiatry*, 24, 145–55.

Wu, J. C., Buchsbaum, M. & Bunney, W. E., Jr. 2001. Clinical neurochemical implications of sleep deprivation's effects on the anterior cingulate of depressed responders. *Neuropsychopharmacology*, 25, S74–S8.

Xu, A., Sakurai, E., Kuramasu, A., Zhang, J., Li, J., Okamura, N., Zhang, D., Yoshikawa, T., Watanabe, T. & Yanai, K. 2010. Roles of hypothalamic subgroup histamine and orexin neurons on behavioral responses to sleep deprivation induced by the treadmill method in adolescent rats. *J Pharmacol Sci*, 114, 444–53.

Xu, W., Zheng, D., Liu, Y., Li, J., Yang, L. & Shang, X. 2017. Glaucocalyxin B alleviates lipopolysaccharide-induced Parkinson's disease by inhibiting TLR/NF-kappaB and activating Nrf2/HO-1 pathway. *Cell Physiol Biochem*, 44, 2091–104.

Yamamoto, M., Kensler, T. W. & Motohashi, H. 2018. The KEAP1-NRF2 system: A thiol-based sensor-effector apparatus for maintaining redox homeostasis. *Physiol Rev*, 98, 1169–203.

Yao, W., Zhang, J. C., Ishima, T., Dong, C., Yang, C., Ren, Q., Ma, M., Han, M., Wu, J., Suganuma, H., Ushida, Y., Yamamoto, M. & Hashimoto, K. 2016. Role of Keap1-Nrf2 signaling in depression and dietary intake of glucoraphanin confers stress resilience in mice. *Sci Rep*, 6, 30659.

Yirmiya, R., Rimmerman, N. & Reshef, R. 2015. Depression as a microglial disease. *Trends Neurosci*, 38, 637–58.

Yu, J. B., Shi, J., Gong, L. R., Dong, S. A., Xu, Y., Zhang, Y., Cao, X. S. & Wu, L. L. 2014. Role of Nrf2/ARE pathway in protective effect of electroacupuncture against endotoxic shock-induced acute lung injury in rabbits. *PLoS One*, 9, e104924.

Zada, D., Bronshtein, I., Lerer-Goldshtein, T., Garini, Y. & Appelbaum, L. 2019. Sleep increases chromosome dynamics to enable reduction of accumulating DNA damage in single neurons. *Nat Commun*, 10, 895.

Zammit, G. K., Weiner, J., Damato, N., Sillup, G. P. & Mcmillan, C. A. 1999. Quality of life in people with insomnia. *Sleep*, 22 Suppl 2, S379–85.

Zhang, J. C., Yao, W., Dong, C., Han, M., Shirayama, Y. & Hashimoto, K. 2018. Keap1-Nrf2 signaling pathway confers resilience versus susceptibility to inescapable electric stress. *Eur Arch Psychiatry Clin Neurosci*, 268, 865–70.

Zhang, J. C., Yao, W. & Hashimoto, K. 2016. Brain-derived neurotrophic factor (BDNF)-TrkB signaling in inflammation-related depression and potential therapeutic targets. *Curr Neuropharmacol*, 14, 721–31.

Zhang, C., Zhang, Y. P., Li, Y. Y., Liu, B. P., Wang, H. Y., Li, K. W., Zhao, S. & Song, C. 2019. Minocycline ameliorates depressive behaviors and neuro-immune dysfunction induced by chronic unpredictable mild stress in the rat. *Behav Brain Res*, 356, 348–57.

Zielinski, M. R. & Krueger, J. M. 2011. Sleep and innate immunity. *Front Biosci (Schol Ed)*, 3, 632–42.

Zola-Morgan, S., Squire, L. R. & Amaral, D. G. 1986. Human amnesia and the medial temporal region: Enduring memory impairment following a bilateral lesion limited to field CA1 of the hippocampus. *J Neurosci*, 6, 2950–67.

6 Artificial Intelligence in Drug Discovery and Repurposing for Neurodegenerative Disorders

Muhammad Liaquat Raza and Muhammad Khizar Abbas

6.1 INTRODUCTION

Neurodegenerative disorders, including Alzheimer's disease (AD) and Parkinson's disease (PD), pose an escalating challenge to global healthcare systems (Ding et al., 2022; Huang et al., 2023; GBD 2016 Neurology Collaborators, 2019). With the aging population, the prevalence of these complex disorders has surged, demanding innovative strategies for understanding, diagnosing, and treating them effectively. This chapter explores the intersection of artificial intelligence (AI) and drug discovery in the context of neurodegenerative disorders, contributing to the broader initiative outlined in the book, "*Emerging Frontiers in Translational Neurodegeneration: Integrating Neuroimaging, Computational Neuroscience, and AI for Diagnosis and Beyond*."

As traditional drug discovery methods face formidable challenges in unraveling the intricate mechanisms underlying neurodegenerative disorders, the need for transformative approaches becomes imperative. AI emerges as a potent ally, capable of deciphering vast datasets and unveiling hidden patterns within the intricate biological landscape (Cummings et al., 2023; Fang et al., 2022; Zagórska et al., 2023).

The pressing nature of this issue is evident when considering the immense socioeconomic burden these disorders impose globally. AD, for instance, affects over 50 million people worldwide, with projections indicating a dramatic increase in the coming decades (Alzheimer's disease facts and figures, 2023). PD, another prominent neurodegenerative disorder, adds to this burden, emphasizing the urgency for novel therapeutic interventions (Cong et al., 2022).

This chapter navigates through the evolving landscape of AI-driven drug discovery, shedding light on its potential to revolutionize our approach to neurodegenerative disorders. By leveraging the power of machine learning and deep learning, AI holds promise in identifying novel drug candidates and repurposing existing ones (Vemula et al., 2023). The integration of AI with neuroimaging and computational neuroscience further enhances the depth and precision of our understanding, paving the way for more effective diagnostic and therapeutic strategies (Sarkar et al., 2023).

In the subsequent sections, we delve into the challenges posed by traditional drug discovery methods, the role of AI in reshaping this landscape, and specific applications of AI in neurodegenerative drug discovery. Through illustrative case studies, we showcase instances where AI has successfully identified potential drugs and repurposed existing ones for these

DOI: 10.1201/9781032711102-6

challenging disorders. The ethical considerations and future prospects of this transformative approach are also explored, offering a comprehensive view of the evolving paradigm in neurodegenerative drug discovery.

6.2 NEURODEGENERATIVE DISORDERS: A BRIEF OVERVIEW

Neurodegenerative disorders represent a diverse group of chronic conditions characterized by the progressive degeneration of the structure and function of the nervous system. Among the myriad of neurodegenerative disorders, AD and PD stand out prominently (Table 6.1), posing significant challenges to healthcare systems globally.

6.2.1 Alzheimer's Disease

AD is the most common form of dementia, affecting memory, cognitive function, and daily activities. The hallmark pathological features of AD include the accumulation of beta-amyloid plaques and tau tangles in the brain. These abnormalities lead to the progressive loss of neurons, particularly in regions associated with memory and learning. As the disease advances, individuals experience cognitive decline, behavioral changes, and ultimately, a profound impact on their quality of life (Ebenau et al., 2020; Erickson et al., 2023; Querry et al., 2023).

The prevalence of AD is staggering, with over 50 million people worldwide affected, and projections indicating a steep rise in the coming decades (Alzheimer's disease facts and figures, 2023). Despite decades of research, the precise etiology of AD remains elusive, presenting a major obstacle in the development of effective therapeutic interventions.

6.2.2 Parkinson's Disease

PD is a neurodegenerative disorder characterized by the degeneration of dopamine-producing neurons in the substantia nigra region of the brain. This loss of dopamine leads to motor symptoms such as tremors, rigidity, and bradykinesia. PD is not only a movement disorder but also affects various non-motor functions, contributing to a wide range of symptoms that impact daily life (Bock et al., 2023; Gökçal et al., 2017; Prajjwal et al., 2023; von der Recke et al., 2023).

TABLE 6.1
Overview of Neurodegenerative Disorders

Disorder	Prevalence	Key Features
Alzheimer's Disease	>50 million	Beta-amyloid plaques, tau tangles
Parkinson's Disease	Millions	Dopamine-producing neuron degeneration
Amyotrophic Lateral Sclerosis	5–7 per 100,000	Motor neuron degeneration
Huntington's Disease	5–10 per 100,000	Mutated huntingtin gene, progressive nerve cell degeneration
Multiple Sclerosis	2.3 million	Autoimmune, demyelination of nerve fibers
Frontotemporal Dementia	<5 per 100,000	Frontal and temporal lobe degeneration

The global burden of PD is substantial, affecting millions of individuals, with an increasing prevalence as populations age. Similar to AD, PD lacks a definitive cure, and current treatments aim at alleviating symptoms rather than addressing the underlying neurodegenerative processes (Laar et al., 2023; Prajjwal et al., 2023).

6.3 CHALLENGES IN TREATING NEURODEGENERATIVE DISORDERS

6.3.1 Complex Pathophysiology

The intricate pathophysiology of neurodegenerative disorders presents a formidable challenge. The interplay of genetic, environmental, and lifestyle factors contributes to the complexity of these conditions (Prajjwal et al., 2023; Zhao et al., 2023). Life factors that affect the PD patients' include sleep (Zhang et al., 2022) and heat exposure (Clark & Rigby, 2023). Factors such as exercise (Flynn et al., 2022) and mindfulness techniques (Kwok et al., 2022) have a positive impact on PD patients. Unraveling the precise mechanisms triggering neuronal degeneration requires a comprehensive understanding of the multifaceted interactions within the brain's intricate networks (Müller-Nedebock et al., 2023; Shi et al., 2023).

6.4 BEYOND AD AND PD: DIVERSE NEURODEGENERATIVE DISORDERS

6.4.1 Amyotrophic Lateral Sclerosis

Amyotrophic lateral sclerosis (ALS), also known as Lou Gehrig's disease, is characterized by the progressive degeneration of motor neurons (Dey et al., 2023). This leads to muscle weakness, paralysis, and, ultimately, respiratory failure (Martin et al., 2017). The exact etiology of ALS is complex, involving genetic and environmental factors. Developing effective treatments for ALS remains a significant challenge, highlighting the urgent need for innovative therapeutic strategies (Tzeplaeff et al., 2023).

6.4.2 Huntington's Disease

Huntington's (HD) disease is a hereditary disorder caused by a mutated huntingtin gene. It leads to the progressive degeneration of nerve cells in the brain, affecting movement, cognition, and psychiatric function (Dai et al., 2023). HD presents unique challenges due to its genetic nature, emphasizing the importance of precision medicine in its management (Teli et al., 2023).

6.4.3 Multiple Sclerosis

Multiple sclerosis (MS) is an autoimmune disorder in which the immune system attacks the protective covering of nerve fibers. This results in communication disruptions between the brain and the rest of the body. The varied clinical presentations and relapsing-remitting nature of MS make its diagnosis and management intricate (Göttle et al., 2024; Jakimovski et al., 2024).

6.4.4 Frontotemporal Dementia

Frontotemporal dementia (FTD) encompasses a group of disorders characterized by degeneration in the frontal and temporal lobes of the brain (Ulugut & Pijnenburg, 2023). It manifests as changes in behavior, personality, and language (Nevler et al., 2024). The heterogeneity of FTD poses diagnostic challenges (Pressman et al., 2024), necessitating a deeper understanding of its underlying molecular mechanisms (Halder & Drummond, 2024).

6.5 CHALLENGES IN UNDERSTANDING DIVERSE NEURODEGENERATIVE DISORDERS

6.5.1 Shared Pathological Complexity

The realm of neurodegenerative disorders is marked by a shared pathological complexity that transcends individual conditions. At the core of this complexity are recurring themes such as protein misfolding, neuroinflammation, and mitochondrial dysfunction. These intertwined molecular processes cut across various disorders, demanding a comprehensive exploration of the interconnected pathways that underlie conditions ranging from AD to PD and beyond. While each neurodegenerative disorder has its unique features, there exists a shared pathological complexity. Protein misfolding, neuroinflammation, and mitochondrial dysfunction are recurring themes, requiring comprehensive investigation to unravel the interconnected molecular pathways (Han et al., 2024; Kopeć et al., 2023; Peng et al., 2020; Soto & Pritzkow, 2018; Teleanu et al., 2022).

Moreover, the role of neuroinflammation cannot be understated. Inflammatory processes within the brain are evident across various neurodegenerative disorders, contributing to disease progression. Microglia, the resident immune cells of the brain, play a central role in this inflammatory response. Understanding the intricate interplay between misfolded proteins and the immune system is crucial for devising strategies that could modulate neuroinflammation and potentially alter the trajectory of these disorders (Peng et al., 2020; Teleanu et al., 2022).

Mitochondrial dysfunction, characterized by impaired energy production and increased oxidative stress, is another recurring theme. The mitochondria, often referred to as the powerhouse of the cell, are crucial for neuronal function. When compromised, they contribute to the overall degeneration of neurons. Exploring the shared pathways leading to mitochondrial dysfunction could unveil potential therapeutic targets with broad applicability across diverse neurodegenerative conditions (Dai et al., 2023).

6.5.2 Diagnostic Heterogeneity

The diagnostic landscape of neurodegenerative disorders is marked by considerable heterogeneity, presenting a formidable challenge in the quest for accurate and early identification. The diagnostic heterogeneity among neurodegenerative disorders adds an additional layer of complexity. Overlapping symptoms and shared pathological markers necessitate precise diagnostic tools and biomarkers to distinguish between conditions accurately (Chen et al., 2024; Feczko et al., 2019). Consider, for instance, the challenges in distinguishing between PD and atypical parkinsonian syndromes. While PD is characterized by the classic motor symptoms of tremors, bradykinesia, and rigidity, atypical parkinsonian syndromes present with overlapping symptoms, making precise diagnosis a daunting task. This diagnostic ambiguity can delay appropriate interventions and contribute to the challenges of conducting accurate clinical trials targeting specific disorders (Kong et al., 2023; Lustoza Rodrigues et al., 2023; Rossetti et al., 2019).

Advancements in neuroimaging and biomarker discovery offer promising avenues for addressing diagnostic heterogeneity. Imaging techniques such as positron emission tomography (PET) and magnetic resonance imaging (MRI) provide detailed insights into structural and functional changes within the brain. Additionally, identifying reliable biomarkers associated with specific disorders could serve as diagnostic beacons, guiding clinicians toward more accurate and timely diagnoses (Sasikumar & Strafella, 2023).

6.5.3 Therapeutic Diversification

The diversification of underlying mechanisms across neurodegenerative disorders poses a substantial hurdle in the development of effective therapeutic interventions. Traditional drug discovery approaches often target specific molecular pathways associated with individual conditions. However, the intricate web of molecular processes shared among various disorders necessitates a more nuanced and diversified approach to treatment strategies. Developing therapeutic interventions becomes challenging due to the diversification of underlying mechanisms across neurodegenerative disorders. Tailored approaches are crucial, considering the genetic, molecular, and clinical differences among these conditions (Aleksandrova & Neganova, 2023; Karagianni et al., 2022).

In AD, the focus often revolves around mitigating the accumulation of beta-amyloid plaques and tau tangles. In PD, the emphasis is on modulating dopamine levels and addressing alpha-synuclein misfolding. Each disorder demands a tailored therapeutic strategy, considering its distinct genetic, molecular, and clinical features. Consequently, the challenge lies in developing interventions that can accommodate this diversity, fostering a paradigm shift from a one-size-fits-all approach to a more personalized and targeted therapeutic landscape (Marras et al., 2024; Yang et al., 2023).

Moreover, the lack of disease-modifying agents further accentuates the therapeutic challenges. Many existing treatments primarily offer symptomatic relief without addressing the underlying causes of neurodegeneration. The quest for disease-modifying treatments involves deciphering the intricacies of each disorder's progression, aiming to halt or slow down the degenerative processes. This requires a profound understanding of the shared and unique mechanisms driving neurodegeneration across diverse conditions (Kong et al., 2023; Lustoza Rodrigues et al., 2023).

6.5.4 Unmet Medical Needs

The landscape of neurodegenerative disorders is marked by substantial unmet medical needs, highlighting the urgency for innovative approaches in drug discovery and development. Despite the prevalence and impact of these disorders on global health, effective treatments that modify disease progression remain elusive for many conditions. Many neurodegenerative disorders lack effective treatments, contributing to the substantial unmet medical needs in this field. Addressing these needs requires a paradigm shift in our understanding and approach to drug discovery and development (Domínguez-Fernández et al., 2023; Kashyap & Siddiqi, 2021; Pasrija et al., 2022).

In the case of ALS, a devastating disorder characterized by progressive motor neuron degeneration, the limited success in developing targeted therapies underscores the urgent need for breakthroughs. The heterogeneity of ALS, with both sporadic and familial forms, adds complexity to treatment strategies. Addressing the unmet medical needs in ALS necessitates unraveling the intricate genetic and molecular factors contributing to its pathogenesis (Tzeplaeff et al., 2023).

Similarly, in the realm of FTD, characterized by degeneration in the frontal and temporal lobes of the brain, the lack of effective treatments reflects the challenges in understanding the underlying molecular mechanisms. The clinical heterogeneity within FTD, presenting as behavioral, language, or motor variants, further complicates therapeutic endeavors. Innovative strategies are imperative to bridge the gap between unmet medical needs and the development of targeted interventions (Pressman et al., 2024; Ulugut & Pijnenburg, 2023).

6.5.5 Recent Advances

Recent advancements in neurodegenerative disorder research extend beyond AD and PD. Genetic studies have unearthed key insights into the molecular basis of ALS and HD (Rani et al., 2023). Imaging techniques, such as functional MRI (fMRI) and diffusion tensor imaging, have enhanced our understanding of MS (Weeda et al., 2024). The identification of potential biomarkers for FTD underscores the progress in diagnostic approaches (Maheshwari & Singh, 2024).

As we navigate through the complexities of diverse neurodegenerative disorders, the integration of AI in drug discovery emerges as a promising avenue. Addressing these unmet medical needs requires a holistic and collaborative approach, bringing together researchers, clinicians, and pharmaceutical industries to explore novel avenues in drug discovery. The advent of AI has emerged as a promising ally in this pursuit, offering innovative solutions to the challenges posed by the shared complexity, diagnostic heterogeneity, therapeutic diversification, and overall unmet medical needs in the realm of neurodegenerative disorders.

In the subsequent sections of this chapter, we delve into how AI serves as a transformative force in overcoming these challenges, reshaping the landscape of neurodegenerative disorder research and treatment. By leveraging advanced computational techniques, machine learning, and neuroimaging, AI provides a comprehensive and data-driven approach that holds the potential to unlock new frontiers in translational neurodegeneration research.

6.5.6 Traditional Drug Discovery Challenges

Traditional drug discovery approaches, while instrumental in bringing numerous medications to market, encounter substantial challenges when tackling the intricate nature of neurodegenerative disorders. The pursuit of novel therapeutics through traditional drug discovery approaches is fraught with an array of challenges, spanning from the initial stages of target identification to the final phases of clinical trials. These challenges have perennially impeded the efficiency of drug development and contributed to the high attrition rates in the pharmaceutical industry. In this section, we delve into the multifaceted obstacles faced by researchers and pharmaceutical companies in the traditional drug discovery paradigm. It is important to emphasize the critical need for innovative strategies in the face of escalating complexities (Butlen-Ducuing et al., 2016).

6.5.7 Target Identification and Validation

One of the primary hurdles in traditional drug discovery for neurodegenerative disorders lies in the identification and validation of suitable drug targets. The foundational step in drug discovery involves identifying and validating a suitable molecular target implicated in the pathogenesis of a particular disease. This initial phase is often beset by the challenge of choosing a target that is both biologically relevant and druggable. Identifying a target that plays a crucial role in the disease process and is amenable to pharmacological modulation is pivotal for the success of subsequent drug development efforts.

The intricate web of molecular pathways involved in these disorders often lacks well-defined and druggable targets. This challenge is particularly pronounced when attempting to modify disease progression rather than merely alleviating symptoms (Loewa et al., 2023; Rossetti et al., 2019; Swalley, 2020).

Compounding this challenge is the limited understanding of the complex molecular networks underlying many diseases. In neurodegenerative disorders, for instance, the intricate interplay of genetic, environmental, and lifestyle factors adds layers of complexity to target identification. Furthermore, the validation of a potential target's role in disease pathogenesis often requires extensive experimental evidence, which can be resource-intensive and time-consuming.

6.5.8 High Attrition Rates

The drug development pipeline faces a high attrition rate, especially during clinical trials. Neurodegenerative disorders exhibit considerable heterogeneity, leading to varied patient responses and complicating the evaluation of drug efficacy. The lengthy and costly nature of clinical trials exacerbates the impact of high attrition rates, necessitating a more streamlined approach (Moore et al., 2020).

6.5.8.1 Hit-to-Lead Optimization

The transition from hits obtained in high-throughput screens to lead compounds suitable for further development is a critical phase in drug discovery. Hit-to-lead optimization aims to refine and enhance the properties of initial hits to improve their drug-like characteristics, such as potency, selectivity, and pharmacokinetic profiles.

However, this process is rife with challenges, including issues related to chemical tractability and compound synthesis. Many initial hits may exhibit poor solubility, bioavailability, or metabolic stability, necessitating extensive medicinal chemistry efforts to optimize their properties. Moreover, achieving the desired balance between potency and selectivity is often a delicate and time-consuming process.

6.5.8.2 Preclinical Models and Predictive Validity

Reliance on preclinical models, such as cell lines and animal models, introduces another layer of complexity and uncertainty into drug discovery. Despite the advances in creating genetically modified animals that more accurately recapitulate human disease conditions, the translational gap between preclinical models and human responses remains a persistent challenge.

The lack of predictive validity of preclinical models can lead to the development of compounds that demonstrate efficacy in animal studies but fail to translate into clinical success. This discordance contributes significantly to the high attrition rates observed during clinical trials and underscores the need for more predictive and relevant models that better mimic human physiology and disease progression.

6.5.9 Blood-Brain Barrier Limitations

The blood-brain barrier (BBB) poses a formidable obstacle in delivering therapeutic agents to the central nervous system. Many potential drugs struggle to cross this protective barrier, restricting their effectiveness in reaching target brain regions. Overcoming BBB limitations is crucial for ensuring that drugs can exert their desired effects within the brain (Wu et al., 2023).

6.5.10 Lack of Disease-Modifying Agents

A prevalent challenge in traditional drug discovery for neurodegenerative disorders is the scarcity of disease-modifying agents. Current treatments predominantly focus on symptom management, offering temporary relief without altering the underlying disease progression. Developing therapies that address the root causes and modify the course of neurodegenerative disorders remains an unmet need (Karagianni et al., 2022; Loewa et al., 2023; Rossetti et al., 2019; Swalley, 2020).

6.5.11 Limited Understanding of Disease Mechanisms

The intricate and multifaceted nature of neurodegenerative disorders contributes to a limited understanding of the underlying disease mechanisms. Traditional drug discovery often relies on specific molecular targets, yet the interconnected pathways and complex etiology of these disorders demand a more holistic approach. A deeper understanding of disease biology is essential for the identification of effective drug targets (Chen et al., 2024; Maheshwari & Singh, 2024; Rossetti et al., 2019).

6.5.12 Lack of Personalization

Neurodegenerative disorders exhibit significant inter-individual variability, requiring personalized approaches for optimal therapeutic outcomes. Traditional drug discovery methods often adopt a one-size-fits-all approach, neglecting the importance of individual genetic, environmental, and lifestyle factors that influence treatment responses. Personalized medicine holds the potential to enhance treatment efficacy and minimize adverse effects (Domínguez-Fernández et al., 2023; Hampel et al., 2023a, 2023b).

6.5.13 Reproducibility Challenges

The issue of irreproducibility is a concern in traditional drug discovery. Published research findings may not always be replicable, leading to delays in drug development and potential misdirection of resources. Addressing reproducibility challenges is crucial for building a robust foundation of knowledge and accelerating the translation of research findings into clinical applications (Butlen-Ducuing et al., 2016; Kesselheim et al., 2015).

6.5.14 Integration of Systems Biology

In the era of precision medicine, the integration of systems biology approaches is becoming increasingly vital. Traditional drug discovery often operates in isolation, overlooking the complex interactions within biological systems. Incorporating systems biology methodologies allows for a more comprehensive understanding of the intricate networks involved in neurodegenerative disorders (Fasano & Alberio, 2023).

6.5.14.1 Regulatory Hurdles and Safety Concerns

Navigating the regulatory landscape poses yet another hurdle in traditional drug discovery. Stringent regulatory requirements demand exhaustive safety and efficacy data, leading to prolonged development timelines and increased costs. Moreover, concerns about potential side effects or unforeseen toxicities can result in the discontinuation of promising drug candidates, further contributing to the attrition rates observed in drug development.

Addressing safety concerns is particularly critical in the context of neurodegenerative disorders, where the central nervous system is the primary target. The BBB and potential off-target effects present unique challenges in ensuring the safety of candidate compounds. Balancing the need for efficacy with stringent safety requirements remains a delicate equilibrium.

6.5.14.2 Economic Pressures and Resource Allocation

The financial burden associated with drug discovery and development is a pervasive challenge. The extensive resources required for target identification, compound screening, preclinical testing, and clinical trials necessitate substantial financial investments. The inherent uncertainty and risk associated with drug development further exacerbate these economic pressures.

The long timelines inherent in traditional drug discovery can strain financial resources, especially for smaller biotech companies. This dynamic often influences strategic decision-making, potentially leading to the premature abandonment of promising compounds or the prioritization of financially safer but less innovative projects.

6.5.14.3 Embracing Technological Advancements

The advancement of technology, particularly in high-throughput screening, omics technologies, and imaging modalities, presents an opportunity to overcome traditional drug discovery challenges. Integration of these technologies enables a more systematic and data-driven approach, paving the way for enhanced target identification, validation, and understanding of disease mechanisms (Alvarado et al., 2024; La Cognata et al., 2021; Ruffini et al., 2020).

In the subsequent sections, we explore how AI serves as a transformative force, addressing these traditional drug discovery challenges and offering innovative solutions to accelerate the development of effective treatments for neurodegenerative disorders.

6.6 ROLE OF AI IN DRUG DISCOVERY

6.6.1 AI and Its Potential in Revolutionizing Drug Discovery

AI has emerged as a transformative force in drug discovery, offering novel approaches to overcome traditional challenges. At its core, AI refers to the development of algorithms and computational models that mimic human intelligence, enabling machines to analyze complex datasets, identify patterns, and make data-driven predictions. In the realm of drug discovery, AI holds the promise of significantly expediting the identification of potential drug candidates, optimizing clinical trial designs, and enhancing the overall efficiency of the drug development process (Brogi & Calderone, 2021; Schneider et al., 2020).

The power of AI lies in its ability to handle vast amounts of biological data with speed and precision. Machine learning and deep learning algorithms, two subsets of AI, are particularly instrumental in this regard. Machine learning algorithms can discern patterns from data without explicit programming, while deep learning, inspired by the structure and function of the human brain, excels in processing intricate and unstructured data, such as genomics, proteomics, and imaging data (Patel et al., 2020; Schneider et al., 2020; Vamathevan et al., 2019).

6.6.2 Machine Learning and Deep Learning in Analyzing Complex Biological Data

The application of AI in drug discovery begins with the analysis of complex biological data. Machine learning algorithms can sift through massive datasets, identifying relevant patterns and associations that might elude traditional methods. This is particularly valuable in the context of neurodegenerative disorders, where the multifaceted nature of the diseases demands a comprehensive understanding of intricate biological pathways (Patel et al., 2020; Schneider et al., 2020; Vamathevan et al., 2019; Yang et al., 2022).

Deep learning, with its ability to decipher intricate patterns in large datasets, proves invaluable in decoding the complexities of genomics and proteomics. For instance, deep learning models can analyze gene expression profiles, identify biomarkers indicative of disease progression, and unveil potential therapeutic targets. Additionally, the integration of imaging data, such as neuroimaging scans, into deep learning models enhances our ability to visualize structural and functional changes in the brain, aiding in early diagnosis and treatment monitoring (Pasrija et al., 2022; Sarkar et al., 2023; Talevi, 2024).

The synergy between AI, machine learning, and deep learning heralds a new era in drug discovery, where data-driven insights facilitate more informed decision-making. As we delve into the specific application of AI in neurodegenerative drug discovery (Table 6.2), the convergence of advanced computational techniques with biological understanding becomes evident.

TABLE 6.2
AI Applications in Neurodegenerative Drug Discovery

Application	Description
Drug Target Identification	AI analyzes biological data to identify potential drug targets
Biomarker Discovery	AI identifies biomarkers indicative of disease progression
Predictive Modeling	AI models predict disease trajectory and treatment responses
Personalized Medicine	Integrating AI with patient data for tailored interventions
Connectivity Analysis	Analyzing neural network connectivity for targeted drug discovery

6.7 AI IN NEURODEGENERATIVE DRUG DISCOVERY

6.7.1 Identifying Potential Drugs for Alzheimer's Disease

AI has demonstrated remarkable success in identifying potential drugs for AD. By analyzing vast datasets encompassing genetic information, protein interactions, and clinical data, machine learning algorithms can uncover novel drug candidates targeting specific pathways implicated in AD pathogenesis. For instance, a study utilizing AI identified a repurposable drug that showed promise in modulating beta-amyloid aggregation, a key pathological feature of AD (Hajjo et al., 2022; Mishra & Li, 2020; Silva-Spínola et al., 2022).

6.7.2 Precision Medicine Approaches in Parkinson's Disease

AI's application extends to personalized medicine in PD. By analyzing diverse datasets, including genetic information, clinical records, and neuroimaging data, AI models can identify subgroups of PD patients with distinct molecular profiles (D'Sa et al., 2023; Sasikumar & Strafella, 2023). This enables the development of tailored therapeutic approaches, optimizing treatment responses and minimizing side effects (Marras et al., 2024).

These case studies underscore the versatility of AI in neurodegenerative drug discovery, showcasing its ability to unravel complex biological interactions, identify potential drug candidates, and tailor treatments to individual patients. The integration of AI into the drug discovery pipeline not only accelerates the identification of novel therapeutics but also enhances the precision and efficacy of interventions, marking a paradigm shift in our approach to neurodegenerative disorders. The subsequent sections delve into the broader landscape of drug repurposing facilitated by AI and the integration of neuroimaging and computational neuroscience in this transformative journey (Horowitz et al., 2022).

6.7.3 Integrating Neuroimaging and Computational Neuroscience

The convergence of AI, neuroimaging, and computational neuroscience represents a groundbreaking approach in comprehending the intricacies of neurodegenerative disorders. This section illuminates the synergistic relationship between these three domains, illustrating how their integration forms a powerful and comprehensive framework for understanding, diagnosing, and treating complex neurological conditions (Renton et al., 2024; Sasikumar & Strafella, 2023).

6.8 SYNERGY BETWEEN AI AND NEUROIMAGING

6.8.1 Enhanced Visualization and Diagnosis

Neuroimaging techniques, including MRI, PET, and fMRI, provide unprecedented insights into the structural and functional alterations within the brain. AI augments these capabilities by precisely analyzing neuroimaging data, enabling enhanced visualization of subtle changes associated with neurodegenerative disorders. This synergy enhances diagnostic accuracy, allowing for early detection and intervention (Horowitz et al., 2022; Kamagata et al., 2021).

6.8.2 Biomarker Discovery and Predictive Modeling

The combination of AI and neuroimaging facilitates the discovery of biomarkers indicative of disease progression. Machine learning algorithms can sift through extensive datasets, identifying subtle patterns that serve as potential biomarkers. These biomarkers not only aid in disease diagnosis but also contribute to predictive modeling, offering insights into the trajectory of neurodegenerative disorders and potential response to treatments (Doroszkiewicz & Groblewska, 2022).

6.8.3 Biomarkers in Drug Discovery

6.8.3.1 Biomarker Discovery and Predictive Modeling in Neurodegenerative Drug Discovery

Biomarker discovery and predictive modeling play pivotal roles in advancing neurodegenerative drug discovery, offering innovative approaches to understand, diagnose, and treat complex disorders such as AD, PD, and other related ailments. In this section, we explore the significance of biomarkers, the challenges in their discovery, and the transformative impact of predictive modeling in the pursuit of effective therapeutic interventions.

6.8.4 Biomarkers: Unveiling Disease Signatures

Biomarkers, objectively measurable indicators of biological processes or responses to therapeutic interventions, hold immense promise in the field of neurodegenerative drug discovery. The identification and validation of robust biomarkers are crucial steps in understanding disease mechanisms, enabling early and accurate diagnosis, and monitoring treatment responses.

6.8.5 Importance of Biomarkers in Neurodegenerative Disorders

In neurodegenerative disorders, where subtle changes in the brain may precede clinical symptoms, biomarkers serve as windows into the underlying pathophysiology. They provide valuable insights into disease progression, allowing for the identification of potential therapeutic targets and the monitoring of treatment efficacy.

One of the hallmarks of AD, for instance, is the accumulation of beta-amyloid plaques in the brain. Biomarkers such as cerebrospinal fluid (CSF) levels of beta-amyloid and PET imaging of amyloid plaques have emerged as valuable tools for early detection and tracking disease progression. Similarly, in PD, biomarkers like alpha-synuclein aggregates in the CSF or neuroimaging markers can aid in diagnosis and prognosis.

6.8.6 Challenges in Biomarker Discovery

Despite their potential, biomarker discovery in neurodegenerative disorders faces significant challenges. The heterogeneity of these disorders, coupled with variations in disease progression among individuals, complicates the identification of universal biomarkers. Additionally, the need for non-invasive, easily accessible markers adds another layer of complexity.

In the quest for reliable biomarkers, researchers must navigate the intricate interplay of genetic, environmental, and lifestyle factors that contribute to disease manifestation. Moreover, the dynamic nature of neurodegenerative disorders requires biomarkers that can capture changes over time, reflecting the evolving pathology.

6.8.7 Multi-Omics Approaches

Advancements in technology have ushered in an era of multi-omics approaches for biomarker discovery. Integrating genomics, transcriptomics, proteomics, and metabolomics data provides a comprehensive view of the molecular landscape associated with neurodegenerative disorders. These multi-dimensional datasets offer a nuanced understanding of disease mechanisms and unveil potential biomarkers with higher specificity and sensitivity.

For example, combining genetic information with protein expression profiles allows researchers to identify genetic variants that influence protein levels, offering insights into the molecular pathways implicated in disease progression. Integrating diverse omics data also facilitates the identification of novel biomarkers that may not be apparent when analyzing individual datasets in isolation.

6.8.8 Predictive Modeling: Navigating the Complexity

Predictive modeling, driven by AI and machine learning, has emerged as a transformative tool in leveraging biomarker data for neurodegenerative drug discovery. These models, capable of discerning intricate patterns within large and complex datasets, hold the potential to predict disease trajectories, identify high-risk individuals, and guide personalized treatment strategies.

6.8.9 Machine Learning in Predictive Modeling of Biomarkers

Machine learning algorithms, particularly supervised learning approaches, excel in training on labeled datasets to predict outcomes based on input features. In the context of neurodegenerative disorders, predictive modeling harnesses the wealth of biomarker data to create models that can forecast disease progression, stratify patient populations, and even predict responses to specific therapeutic interventions.

6.8.10 Disease Progression Prediction

Predicting the trajectory of neurodegenerative diseases is a complex endeavor due to the multifactorial nature of these disorders. Machine learning models trained on longitudinal biomarker data can, however, capture subtle changes indicative of disease progression. For instance, in AD research, models have been developed to predict cognitive decline based on a combination of genetic, imaging, and clinical data. These predictions offer valuable insights into disease dynamics, aiding in the identification of critical periods for intervention.

6.8.11 Risk Stratification

Identifying individuals at higher risk of developing neurodegenerative disorders is a key application of predictive modeling. Integrating genetic, imaging, and clinical biomarkers, machine learning models can stratify populations based on their likelihood of developing conditions like AD or PD. Early identification of high-risk individuals enables targeted interventions and, potentially, the implementation of preventive measures.

6.8.12 Challenges in Predictive Modeling of Biomarkers

Despite the promise of predictive modeling, several challenges must be addressed to ensure the accuracy and generalizability of these approaches. Overfitting, a phenomenon where models perform well on training data but fail to generalize to new datasets, is a common concern. The dynamic nature of neurodegenerative disorders and the limited availability of longitudinal biomarker data for model training add complexity to this challenge.

Additionally, the interpretability of machine learning models remains a crucial consideration. Understanding the rationale behind model predictions is essential for gaining insights into disease mechanisms and fostering trust among clinicians and researchers. Striking a balance between predictive accuracy and interpretability is an ongoing area of research in the field of predictive modeling for neurodegenerative disorders.

6.8.13 Integration of Biomarkers and Predictive Modeling

The true potential of biomarkers and predictive modeling in neurodegenerative drug discovery is realized when these two approaches synergize. Biomarkers offer a window into the molecular intricacies of disease, while predictive modeling transforms these insights into actionable predictions for diagnosis, prognosis, and treatment.

6.8.14 Early Diagnosis and Intervention

The integration of biomarkers and predictive modeling holds promise for early diagnosis and intervention in neurodegenerative disorders. Biomarkers, indicative of pre-symptomatic changes, can be incorporated into predictive models to identify individuals at high risk of developing the disease. Early intervention strategies, whether pharmacological or lifestyle-based, can then be targeted to this high-risk cohort, potentially delaying or preventing the onset of clinical symptoms.

6.8.14.1 Personalized Treatment Strategies

The heterogeneity observed in neurodegenerative disorders necessitates a personalized approach to treatment. Predictive models, informed by a patient's biomarker profile, can guide the selection of tailored therapeutic interventions. For example, in PD, where response to medication may vary widely among individuals, predictive modeling can help identify patients likely to benefit from specific medications or interventions.

6.8.15 Drug Development and Clinical Trials

Biomarkers play a crucial role in drug development, aiding in the identification of suitable targets and the assessment of treatment efficacy. Predictive modeling enhances this process by predicting the likely success of drug candidates based on preclinical and early clinical data. By identifying potential responders and stratifying patient populations, predictive models contribute to the design of more targeted and efficient clinical trials.

While the integration of biomarkers and predictive modeling holds immense potential, several challenges and considerations must be addressed for these approaches to translate effectively into clinical practice.

6.8.16 Standardization and Reproducibility

Ensuring the standardization and reproducibility of biomarker measurements is critical for the reliability of predictive models. Variability in data collection methods, assay techniques, and data preprocessing steps can introduce noise and hinders the generalizability of models. Establishing standardized protocols and collaborating on large-scale initiatives for data sharing and validation are essential steps in addressing this challenge.

6.8.17 Data Privacy Concern in Biomarker Research

The use of biomarkers, especially in predictive models, raises ethical considerations regarding consent and use of data; therefore, it is a vital concern to ensure the privacy of data in order to avoid any ethical allegation or lawsuit.

6.9 COMPUTATIONAL NEUROSCIENCE'S ROLE IN DECIPHERING BRAIN FUNCTION

6.9.1 Modeling Neural Networks

Computational neuroscience plays a crucial role in developing models that simulate the intricate functioning of neural networks. These models, informed by AI algorithms, can simulate the impact of neurodegenerative processes on brain function. By integrating empirical neuroimaging data, these simulations enhance our understanding of how structural changes translate into functional alterations, providing a more holistic view of the disease mechanisms (Renton et al., 2024; Sasikumar & Strafella, 2023).

6.9.2 Unraveling Network Dynamics

Neurodegenerative disorders often involve disruptions in neural network dynamics. Computational neuroscience, in tandem with AI-driven analyses of neuroimaging data, allows for the unraveling of these complex network dynamics. Understanding how neural networks evolve over the course of the disease is vital for identifying potential intervention points and designing targeted therapeutic strategies (Renton et al., 2024; Sasikumar & Strafella, 2023).

6.10 COMPREHENSIVE APPROACH FOR NEURODEGENERATIVE DISORDERS

6.10.1 Personalized Medicine through Multimodal Data Integration

The integration of AI, neuroimaging, and computational neuroscience facilitates a personalized medicine approach. By analyzing multimodal datasets—combining genetic information, clinical records, neuroimaging scans, and computational models—this comprehensive approach tailors interventions to individual patients. It recognizes the heterogeneity within neurodegenerative disorders, acknowledging that a one-size-fits-all strategy may not be optimal (Domínguez-Fernández et al., 2023).

6.10.2 Drug Discovery Informed by Brain Connectivity

Understanding the connectivity patterns within the brain is pivotal for targeted drug discovery. AI algorithms, informed by neuroimaging and computational neuroscience, can identify specific brain regions or circuits affected by neurodegenerative disorders. This knowledge guides the development of drugs that precisely target these regions, potentially modifying disease progression (Son & Kim, 2024).

6.10.3 Advancing Therapeutic Strategies

The synergy between AI, neuroimaging, and computational neuroscience is transformative in advancing therapeutic strategies. From optimizing drug delivery routes based on neuroimaging insights to designing neuromodulation techniques informed by computational models, this integrated approach opens new avenues for effective and targeted interventions in neurodegenerative disorders (Jin et al., 2022; Kamagata et al., 2021; Silva-Spínola et al., 2022; Talevi, 2024).

The integration of AI, neuroimaging, and computational neuroscience forms a powerful triad, offering a holistic and data-driven approach to understanding and addressing neurodegenerative disorders. As we explore the challenges and ethical considerations in implementing AI in drug discovery, the potential of this multidisciplinary approach becomes increasingly evident, marking a paradigm shift in the landscape of translational neurodegeneration research.

6.11 CHALLENGES AND ETHICAL CONSIDERATIONS

6.11.1 Challenges Faced in Implementing AI in Drug Discovery

Implementing AI in drug discovery, although promising, comes with its set of challenges. One significant hurdle is the need for large and diverse datasets. AI models thrive on data, and in drug discovery, obtaining comprehensive datasets that encapsulate the complexity of neurodegenerative disorders can be challenging (Kong et al., 2023; Lustoza Rodrigues et al., 2023). Additionally, ensuring the quality and representativeness of the data is paramount for reliable AI-driven insights (Jiménez-Luna et al., 2021; Vatansever et al., 2021; Yang et al., 2023).

Interpretable AI is another challenge. As AI algorithms become increasingly complex, understanding how they arrive at specific conclusions can be elusive. In drug discovery, transparency is crucial for gaining trust in the decision-making process. The interpretability of AI models is an ongoing area of research to make their predictions more understandable for researchers and clinicians (David et al., 2019; Jiménez-Luna et al., 2021; Vatansever et al., 2021; Wang et al., 2023).

Integration into existing workflows poses practical challenges. Incorporating AI seamlessly into the drug discovery pipeline requires alignment with traditional methodologies and regulatory frameworks (Ramezani et al., 2023). Bridging the gap between AI innovations and established practices is crucial for fostering widespread adoption.

6.11.2 Addressing Ethical Concerns Surrounding AI Applications in Healthcare

The ethical considerations in deploying AI for neurodegenerative drug discovery are multifaceted. Ensuring data privacy and patient confidentiality is paramount, especially when dealing with sensitive medical information. Safeguarding against biases in AI algorithms is crucial to prevent unintentional discrimination, ensuring fair and equitable outcomes for diverse patient populations (Jiménez-Luna et al., 2021).

Informed consent and patient autonomy are critical aspects when utilizing AI in drug discovery. Patients should be adequately informed about the involvement of AI in their diagnosis and treatment decisions. Respecting patient choices and ensuring transparency in the role of AI technologies upholds ethical standards (Monteith et al., 2023).

6.11.3 Ethical Considerations and Bias Mitigation in AI-Driven Drug Discovery

As AI increasingly becomes an integral part of drug discovery, ethical considerations and the mitigation of biases take center stage. The transformative power of AI in this domain is undeniable, yet it brings forth a set of ethical challenges that demand careful consideration. From ensuring patient privacy to addressing algorithmic biases, ethical frameworks must evolve alongside the technological advancements in AI. In this section, we delve into the ethical considerations surrounding AI in drug discovery and explore strategies for mitigating biases.

6.11.4 Patient Privacy and Data Security

AI-driven drug discovery relies heavily on access to vast and diverse datasets, often including sensitive patient information. Ensuring patient privacy and maintaining the security of health data are paramount ethical considerations in this context. The integration of genomic data, clinical records, and other personal information into AI models poses potential risks, making it imperative to establish robust safeguards.

One key challenge lies in striking a balance between utilizing valuable data for scientific progress and protecting the privacy of individuals. Anonymization and aggregation techniques are commonly employed to de-identify data, but the evolving landscape of re-identification methods necessitates ongoing vigilance. Ethical frameworks must prioritize the establishment of clear guidelines on data anonymization, encryption, and secure data storage practices to safeguard patient privacy throughout the entire drug discovery process.

Additionally, transparency regarding data usage and obtaining informed consent from patients for the use of their data in AI-driven research is essential. Open communication and engagement with the public regarding the ethical implications of AI in drug discovery build trust and foster a sense of accountability within the scientific community.

6.11.5 Informed Consent and Transparency

Informed consent is a cornerstone of ethical medical research, ensuring that individuals understand the nature of the research they are participating in and providing voluntary agreement. In the context of AI-driven drug discovery, challenges arise in articulating the implications of advanced algorithms to study participants. The complex nature of AI models, especially deep learning networks, makes it challenging to convey the intricacies of data processing and potential outcomes to participants with varying levels of technical literacy.

Researchers and institutions must prioritize efforts to enhance transparency and provide accessible information about AI methodologies to study participants. Clear and comprehensible communication regarding the use of AI in data analysis, potential implications for drug discovery, and the intended benefits for healthcare should be incorporated into the informed consent process. This transparency empowers individuals to make informed decisions about participating in AI-driven research.

6.11.6 Algorithmic Bias and Fair Representation

The risk of algorithmic bias is a pervasive concern in AI applications, including drug discovery. Biases can emerge at various stages, from the training data used to develop AI models to the inherent biases present in historical healthcare data. If not carefully addressed, these biases can perpetuate and exacerbate health disparities, leading to inequitable outcomes in drug development and healthcare delivery.

To mitigate algorithmic bias, researchers must critically evaluate training datasets for representativeness and actively work to rectify imbalances. Incorporating diverse datasets that encompass various demographic groups ensures that AI models generalize well across different populations. Additionally, ongoing monitoring and auditing of algorithms for potential biases during their deployment help identify and rectify any disparities that may arise.

Fair representation in AI-driven drug discovery extends beyond demographic factors to include considerations for underrepresented diseases or conditions. Ensuring that AI models adequately account for diverse medical scenarios and rare diseases contributes to more inclusive and equitable drug discovery outcomes.

6.11.7 Explainability and Interpretability

The "black-box" nature of certain AI algorithms, particularly deep learning models, presents challenges in terms of explainability and interpretability. Understanding how AI arrives at specific predictions or recommendations is crucial for both ethical and practical reasons. The lack of interpretability can undermine trust in AI systems, hindering their acceptance within the scientific community and among healthcare practitioners.

Researchers and developers must prioritize efforts to enhance the explainability of AI models, providing transparent insights into decision-making processes. Techniques such as interpretable machine learning algorithms and attention mechanisms offer avenues to demystify complex AI outputs. Emphasizing model interpretability not only aligns with ethical principles but also facilitates the integration of AI-driven insights into the broader healthcare ecosystem.

6.11.8 Accountability and Responsible AI Governance

Establishing accountability frameworks and responsible AI governance structures is essential to navigate the ethical considerations of AI-driven drug discovery. Researchers, institutions, and AI developers should adopt clear policies and guidelines that outline ethical principles, data usage practices, and mechanisms for addressing unforeseen ethical challenges that may arise.

In the realm of AI governance, multidisciplinary collaboration is key. Ethicists, data scientists, clinicians, and legal experts should collaborate to develop comprehensive frameworks that consider the societal, legal, and medical implications of AI applications in drug discovery. This collaborative approach ensures a well-rounded perspective that addresses ethical considerations from various angles.

Moreover, incorporating external oversight mechanisms, such as ethical review boards or regulatory bodies, adds an additional layer of accountability. External reviews provide independent evaluations of AI applications, ensuring that ethical standards are upheld and potential biases are identified and rectified.

6.11.9 Bias Mitigation Strategies

Addressing biases in AI-driven drug discovery requires a proactive and multifaceted approach. Implementing robust bias mitigation strategies helps minimize the impact of biases on research outcomes and ensures equitable representation. Some key strategies include:

- ***Diverse Dataset Curation*:** Actively curating diverse datasets that encompass various demographic groups and disease scenarios is foundational to mitigating biases. Ensuring representative data helps AI models generalize well across different populations.
- ***Bias Auditing and Monitoring*:** Regularly auditing and monitoring AI algorithms for potential biases during both training and deployment phases is crucial. Implementing automated tools to detect and rectify biases in real time contributes to ongoing bias mitigation efforts.
- ***Interdisciplinary Collaboration*:** Collaboration among researchers, data scientists, clinicians, and ethicists fosters a holistic approach to bias mitigation. By bringing together diverse expertise, teams can identify and address biases from different perspectives.
- ***Explainable AI Models*:** Prioritizing the development and utilization of explainable AI models enhances transparency and helps uncover potential biases. Models that provide interpretable outputs empower researchers to understand and rectify biased predictions.
- ***Bias Impact Assessment*:** Conducting impact assessments to evaluate the real-world implications of potential biases is essential. Understanding how biases may affect different populations or diseases guides targeted interventions to minimize disparate outcomes.
- ***Continuous Education and Training*:** Ensuring that researchers and AI practitioners receive ongoing education and training on bias mitigation strategies and ethical considerations promotes a culture of awareness and responsibility.
- ***Community Engagement*:** Actively engaging with communities affected by AI-driven drug discovery ensures that diverse perspectives are considered. Community input can inform research priorities, data collection practices, and ethical guidelines.

Addressing ethical considerations and mitigating biases in AI-driven drug discovery requires a proactive, collaborative, and ongoing effort. By prioritizing patient privacy, enhancing transparency, mitigating algorithmic biases, promoting explainability, establishing accountability frameworks, and implementing bias mitigation strategies, the scientific community can harness the transformative potential of AI while upholding ethical principles and ensuring equitable outcomes in drug development. As AI continues to evolve in the realm of drug discovery for neurodegenerative disorders, navigating these challenges and addressing ethical considerations will be pivotal for fostering a responsible and trustworthy integration of AI into healthcare practices.

6.12 FUTURE PROSPECTS

6.12.1 Exploring Potential Advancements and Future Directions in AI-Driven Drug Discovery for Neurodegenerative Disorders

The future of AI-driven drug discovery for neurodegenerative disorders holds exciting possibilities. Advanced AI algorithms will likely continue to evolve, becoming more adept at handling diverse and complex datasets. This evolution could enable more accurate predictions of disease mechanisms, identification of novel drug targets, and personalized treatment strategies.

The integration of AI with other cutting-edge technologies, such as quantum computing, could further amplify computational capabilities. Quantum computing's ability to process vast amounts of information simultaneously may unlock new frontiers in simulating complex biological processes, expediting drug discovery timelines (Schwalbe & Wahl, 2020).

Collaborations between researchers, pharmaceutical companies, and technology developers will play a pivotal role. Open sharing of data and collaborative initiatives could accelerate progress by creating large, diverse datasets that fuel AI models with deeper insights into the intricacies of neurodegenerative disorders.

In the future, we anticipate the refinement of AI models for predictive diagnostics. These models may not only aid in early disease detection but also forecast disease progression, allowing for more proactive and personalized treatment approaches (Askin et al., 2023). This shift toward preventive and precision medicine holds the potential to revolutionize the management of neurodegenerative disorders.

6.13 CONCLUSION

The integration of AI into neurodegenerative drug discovery represents a transformative paradigm shift in our approach to these complex disorders. From overcoming traditional drug discovery challenges to leveraging neuroimaging and computational neuroscience, AI's impact is profound. As we navigate challenges and ethical considerations, the potential for advancing therapeutic strategies and improving patient outcomes is evident.

Looking ahead, the future of AI-driven drug discovery is filled with promise. With ongoing advancements, interdisciplinary collaborations, and a commitment to ethical deployment, AI has the potential to usher in an era of personalized, targeted interventions for neurodegenerative disorders. As we stand on the cusp of this transformative journey, the synergy between human ingenuity and AI holds the key to unlocking new frontiers in translational neurodegeneration research.

REFERENCES

Alzheimer's disease facts and figures. Alzheimers Dement. 2023 Apr;19(4):1598–1695. doi: 10.1002/alz.13016.

Aleksandrova Y, Neganova M. Deciphering the mysterious relationship between the cross-pathogenetic mechanisms of neurodegenerative and oncological diseases. Int J Mol Sci. 2023 Sep 29;24(19):14766. doi: 10.3390/ijms241914766.

Alvarado CX, Makarious MB, Weller CA, Vitale D, Koretsky MJ, Bandres-Ciga S, Iwaki H, Levine K, Singleton A, Faghri F, Nalls MA, Leonard HL. omicSynth: An open multi-omic community resource for identifying druggable targets across neurodegenerative diseases. Am J Hum Genet. 2024 Jan 4;111(1):150–164. doi: 10.1016/j.ajhg.2023.12.006.

Askin S, Burkhalter D, Calado G, El Dakrouni S. Artificial intelligence applied to clinical trials: Opportunities and challenges. Health Technol (Berl). 2023;13(2):203–213. doi: 10.1007/s12553-023-00738-2.

Bock MA, Vittinghoff E, Bahorik AL, Leng Y, Fink H, Yaffe K. Cognitive and functional trajectories in older adults with prediagnostic Parkinson disease. Neurology. 2023 Mar 28;100(13):e1386–e1394. doi: 10.1212/WNL.0000000000206762.

Brogi, S., Calderone, V. Artificial intelligence in translational medicine. Int. J. Transl. Med. 2021, 1, 223–285.

Butlen-Ducuing F, Pétavy F, Guizzaro L, Zienowicz M, Haas M, Alteri E, Salmonson T, Haas M, Corruble E, Alteri E. Challenges in drug development for central nervous system disorders: A European Medicines Agency perspective. Nat Rev Drug Discov. 2016 Nov 29;15(12):813–814. doi: 10.1038/nrd.2016.237.

Chen P, Zhang S, Zhao K, Kang X, Rittman T, Liu Y. Robustly uncovering the heterogeneity of neurodegenerative disease by using data driven subtyping in neuroimaging: A review. Brain Res. 2024 Jan 15;1823:148675. doi: 10.1016/j.brainres.2023.148675.

Clark CE, Rigby BR. Can exposure to heat attenuate neurodegeneration in older adults with Parkinson's disease? Front Aging Neurosci. 2023 Sep 7;15:1239656. doi: 10.3389/fnagi.2023.1239656.

Cong S, Xiang C, Zhang S, Zhang T, Wang H, Cong S. Prevalence and clinical aspects of depression in Parkinson's disease: A systematic review and meta-analysis of 129 studies. Neurosci Biobehav Rev. 2022 Oct;141:104749. doi: 10.1016/j.neubiorev.2022.104749.

Cummings JL, Gonzalez MI, Pritchard MC, May PC, Toledo-Sherman LM, Harris GA. The therapeutic landscape of tauopathies: Challenges and prospects. Alzheimers Res Ther. 2023 Oct 6;15(1):168. doi: 10.1186/s13195-023-01321-7.

Dai Y, Wang H, Lian A, Li J, Zhao G, Hu S, Li B. A comprehensive perspective of Huntington's disease and mitochondrial dysfunction. Mitochondrion. 2023 May;70:8–19. doi: 10.1016/j.mito.2023.03.001.

David L, Arús-Pous J, Karlsson J, Engkvist O, Bjerrum EJ, Kogej T, Kriegl JM, Beck B, Chen H. Applications of deep-learning in exploiting large-scale and heterogeneous compound data in industrial pharmaceutical research. Front Pharmacol. 2019 Nov 5;10:1303. doi: 10.3389/fphar.2019.01303.

Dey B, Kumar A, Patel AB. Pathomechanistic networks of motor system injury in amyotrophic lateral sclerosis. Curr Neuropharmacol. 2023 Aug 24;22(11):1778–1806. doi: 10.2174/1570159X21666230824091601.

Ding C, Wu Y, Chen X, Chen Y, Wu Z, Lin Z, Kang D, Fang W, Chen F. Global, regional, and national burden and attributable risk factors of neurological disorders: The Global Burden of Disease study 1990-2019. Front Public Health. 2022 Nov 29;10:952161. doi: 10.3389/fpubh.2022.952161.

Domínguez-Fernández C, Egiguren-Ortiz J, Razquin J, Gómez-Galán M, De Las Heras-García L, Paredes-Rodríguez E, Astigarraga E, Miguélez C, Barreda-Gómez G. Review of technological challenges in personalised medicine and early diagnosis of neurodegenerative disorders. Int J Mol Sci. 2023 Feb 7;24(4):3321. doi: 10.3390/ijms24043321.

Doroszkiewicz J., Groblewska M. Molecular biomarkers and their implications for the early diagnosis of selected neurodegenerative diseases. Int. J. Mol. Sci. 2022;23:4610. doi: 10.3390/ijms23094610.

D'Sa K, Evans JR, Virdi GS, Vecchi G, Adam A, Bertolli O, Fleming J, Chang H, Leighton C, Horrocks MH, Athauda D, Choi ML, Gandhi S. Prediction of mechanistic subtypes of Parkinson's using patient-derived stem cell models. Nat Mach Intell. 2023;5(8):933–946. doi: 10.1038/s42256-023-00702-9.

Ebenau JL, Timmers T, Wesselman LMP, et al. ATN classification and clinical progression in subjective cognitive decline: The SCIENCe project. Neurology. 2020;95(1):e46–e58. doi: 10.1212/WNL.0000000000009724.

Erickson P, Simrén J, Brum WS, Ennis GE, Kollmorgen G, Suridjan I, Langhough R, Jonaitis EM, Van Hulle CA, Betthauser TJ, Carlsson CM, Asthana S, Ashton NJ, Johnson SC, Shaw LM, Blennow K, Andreasson U, Bendlin BB, Zetterberg H, ADNI Cohort. Prevalence and clinical implications of a β-amyloid-negative, tau-positive cerebrospinal fluid biomarker profile in Alzheimer disease. JAMA Neurol. 2023 Jul 31;80(9):969–979. doi: 10.1001/jamaneurol.2023.2338.

Fang J, Zhang P, Wang Q, Chiang CW, Zhou Y, Hou Y, Xu J, Chen R, Zhang B, Lewis SJ, Leverenz JB, Pieper AA, Li B, Li L, Cummings J, Cheng F. Artificial intelligence framework identifies candidate targets for drug repurposing in Alzheimer's disease. Alzheimers Res Ther. 2022 Jan 10;14(1):7. doi: 10.1186/s13195-021-00951-z.

Fasano M, Alberio T. Neurodegenerative disorders: From clinicopathology convergence to systems biology divergence. Handb Clin Neurol. 2023;192:73–86. doi: 10.1016/B978-0-323-85538-9.00007-9. PMID: 36796949.

Feczko E, Miranda-Dominguez O, Marr M, Graham AM, Nigg JT, Fair DA. The heterogeneity problem: Approaches to identify psychiatric subtypes. Trends Cogn Sci. 2019 Jul;23(7):584–601. doi: 10.1016/j.tics.2019.03.009.

Flynn A, Dennis S, Preston E, Canning CG, Allen NE. Exercising with Parkinson's: The good, the bad and the need for support to keep exercising. A qualitative study. Clin Rehabil. 2022 Oct;36(10):1332–1341. doi: 10.1177/02692155221100884.

GBD 2016 Neurology Collaborators. Global, regional, and national burden of neurological disorders, 1990–2016: A systematic analysis for the Global Burden of Disease Study 2016. Lancet Neurol. 2019 May;18(5):459–480. doi: 10.1016/S1474-4422(18)30499-X.

Gökçal E, Gür VE, Selvitop R, Babacan Yildiz G, Asil T. Motor and non-motor symptoms in Parkinson's disease: Effects on quality of life. Noro Psikiyatr Ars. 2017 Jun;54(2):143–148.

Göttle P, Dietrich M, Küry P. Multiple sclerosis drug repurposing for neuroregeneration. Neural Regen Res. 2024 Mar;19(3):507–508. doi: 10.4103/1673-5374.380901.

Hajjo R, Sabbah DA, Abusara OH, Al Bawab AQ. A review of the recent advances in Alzheimer's disease research and the utilization of network biology approaches for prioritizing diagnostics and therapeutics. Diagnostics (Basel). 2022 Nov 28;12(12):2975. doi: 10.3390/diagnostics12122975.

Halder A, Drummond E. Strategies for translating proteomics discoveries into drug discovery for dementia. Neural Regen Res. 2024 Jan;19(1):132–139. doi: 10.4103/1673-5374.373681.

Hampel H, Caruso G, Nisticò R, Piccioni G, Mercuri NB, Giorgi FS, Ferrarelli F, Lemercier P, Caraci F, Lista S, Vergallo A. Neurodegeneration Precision Medicine Initiative Npmi. Biological mechanism-based neurology and psychiatry: A BACE1/2 and downstream pathway model. Curr Neuropharmacol. 2023b;21(1):31–53. doi: 10.2174/1570159X19666211201095701.

Hampel H, Gao P, Cummings J, Toschi N, Thompson PM, Hu Y, Cho M, Vergallo A. The foundation and architecture of precision medicine in neurology and psychiatry. Trends Neurosci. 2023a;46(3):176–198. doi: 10.1016/j.tins.2022.12.004.

Han T, Xu Y, Sun L, Hashimoto M, Wei J. Microglial response to aging and neuroinflammation in the development of neurodegenerative diseases. Neural Regen Res. 2024 Jun 1;19(6):1241–1248. doi: 10.4103/1673-5374.385845.

Horowitz T, Grimaldi S, Azulay JP, Guedj E. Molecular imaging in Parkinsonism: The essential for clinical practice and future perspectives. Rev Neurol (Paris). 2022 May;178(5):484–489. doi: 10.1016/j.neurol.2022.03.002.

Huang Y, Li Y, Pan H, Han L. Global, regional, and national burden of neurological disorders in 204 countries and territories worldwide. J Glob Health. 2023 Nov 29;13:04160. doi: 10.7189/jogh.13.04160.

Jakimovski D, Bittner S, Zivadinov R, Morrow SA, Benedict RH, Zipp F, Weinstock-Guttman B. Multiple sclerosis. Lancet. 2024 Jan 13;403(10422):183–202. doi: 10.1016/S0140-6736(23)01473-3.

Jiménez-Luna J, Grisoni F, Weskamp N, Schneider G. Artificial intelligence in drug discovery: Recent advances and future perspectives. Expert Opin Drug Discov. 2021 Sep;16(9):949–959. doi: 10.1080/17460441.2021.1909567.

Jin S, Lv Z, Kang L, Wang J, Tan C, Shen L, Wang L, Liu J. Next generation of neurological therapeutics: Native and bioengineered extracellular vesicles derived from stem cells. Asian J Pharm Sci. 2022 Nov;17(6):779–797. doi: 10.1016/j.ajps.2022.10.002.

Kamagata K, Andica C, Kato A, Saito Y, Uchida W, Hatano T, Lukies M, Ogawa T, Takeshige-Amano H, Akashi T, Hagiwara A, Fujita S, Aoki S. Diffusion magnetic resonance imaging-based biomarkers for neurodegenerative diseases. Int J Mol Sci. 2021 May 14;22(10):5216. doi: 10.3390/ijms22105216.

Karagianni K, Pettas S, Christoforidou G, Kanata E, Bekas N, Xanthopoulos K, Dafou D, Sklaviadis T. A systematic review of common and brain-disease-specific RNA editing alterations providing novel insights into neurological and neurodegenerative disease manifestations. Biomolecules. 2022 Mar 17;12(3):465. doi: 10.3390/biom12030465.

Kashyap K, Siddiqi MI. Recent trends in artificial intelligence-driven identification and development of anti-neurodegenerative therapeutic agents. Mol Divers. 2021 Aug;25(3):1517–1539. doi: 10.1007/s11030-021-10274-8.

Kesselheim AS, Hwang TJ, Franklin JM. Two decades of new drug development for central nervous system disorders. Nat Rev Drug Discov. 2015 Dec;14(12):815–6. doi: 10.1038/nrd4793.

Kong AH, Wu AJ, Ho OK, Leung MM, Huang AS, Yu Y, Zhang G, Lyu A, Li M, Cheung KH. Exploring the potential of aptamers in targeting neuroinflammation and neurodegenerative disorders: Opportunities and challenges. Int J Mol Sci. 2023 Jul 22;24(14):11780. doi: 10.3390/ijms241411780.

Kopeć K, Szleszkowski S, Koziorowski D, Szlufik S. Glymphatic system and mitochondrial dysfunction as two crucial players in pathophysiology of neurodegenerative disorders. Int J Mol Sci. 2023 Jun 20;24(12):10366. doi: 10.3390/ijms241210366.

Kwok JYY, Lee JJ, Choi EPH, Chau PH, Auyeung M. Stay mindfully active during the coronavirus pandemic: A feasibility study of mHealth-delivered mindfulness yoga program for people with Parkinson's disease. BMC Complement Med Ther. 2022 Feb 7;22(1):37. doi: 10.1186/s12906-022-03519-y.

La Cognata V, Morello G, Cavallaro S. Omics data and their integrative analysis to support stratified medicine in neurodegenerative diseases. Int J Mol Sci. 2021 May 1;22(9):4820. doi: 10.3390/ijms22094820.

Laar A, Silva de Lima AL, Maas BR, Bloem BR, de Vries NM. Successful implementation of technology in the management of Parkinson's disease: Barriers and facilitators. Clin Park Relat Disord. 2023 Feb 16;8:100188. doi: 10.1016/j.prdoa.2023.100188.

Loewa A, Feng JJ, Hedtrich S. Human disease models in drug development. Nat Rev Bioeng. 2023 May 11:1–15. doi: 10.1038/s44222-023-00063-3.

Lustoza Rodrigues TCM, de Sousa NF, Dos Santos AMF, Aires Guimarães RD, Scotti MT, Scotti L. Challenges and discoveries in polypharmacology of neurodegenerative diseases. Curr Top Med Chem. 2023;23(5):349–370. doi: 10.2174/1568026623666230126112628.

Maheshwari S, Singh A. Navigating the dementia landscape: Biomarkers and emerging therapies. Ageing Res Rev. 2024 Jan 10:102193. doi: 10.1016/j.arr.2024.102193.

Marras C, Fereshtehnejad SM, Berg D, Bohnen NI, Dujardin K, Erro R, Espay AJ, Halliday G, Van Hilten JJ, Hu MT, Jeon B, Klein C, Leentjens AFG, Mollenhauer B, Postuma RB, Rodríguez-Violante M, Simuni T, Weintraub D, Lawton M, Mestre TA. Transitioning from subtyping to precision medicine in Parkinson's disease: A purpose-driven approach. Mov Disord. 2024 Jan 20. doi: 10.1002/mds.29708.

Martin S, Al Khleifat A, Al-Chalabi A. What causes amyotrophic lateral sclerosis? F1000Res. 2017 Mar 28;6:371. doi: 10.12688/f1000research.10476.1.

Mishra R, Li B. The application of artificial intelligence in the genetic study of Alzheimer's disease. Aging Dis. 2020 Dec 1;11(6):1567–1584. doi: 10.14336/AD.2020.0312.

Monteith S, Glenn T, Geddes JR, Achtyes ED, Whybrow PC, Bauer M. Challenges and ethical considerations to successfully implement artificial intelligence in clinical medicine and neuroscience: A narrative review. Pharmacopsychiatry. 2023 Nov;56(6):209–213. doi: 10.1055/a-2142-9325.

Moore TJ, Heyward J, Anderson G, Alexander GC. Variation in the estimated costs of pivotal clinical benefit trials supporting the US approval of new therapeutic agents, 2015–2017: A cross-sectional study. BMJ Open. 2020 Jun 11;10(6):e038863. doi: 10.1136/bmjopen-2020-038863.

Müller-Nedebock AC, Dekker MCJ, Farrer MJ, Hattori N, Lim SY, Mellick GD, Rektorová I, Salama M, Schuh AFS, Stoessl AJ, Sue CM, Tan AH, Vidal RL, Klein C, Bardien S. Different pieces of the same puzzle: A multifaceted perspective on the complex biological basis of Parkinson's disease. NPJ Parkinsons Dis. 2023 Jul 13;9(1):110. doi: 10.1038/s41531-023-00535-8.

Nevler N, Cho S, Cousins KAQ, Ash S, Olm CA, Shellikeri S, Agmon G, Gonzalez-Recober C, Xie SX, Barker MS, Manoochehri M, Mcmillan CT, Irwin DJ, Massimo L, Dratch L, Cheran G, Huey ED, Cosentino SA, Van Deerlin VM, Liberman MY, Grossman M. Changes in digital speech measures in asymptomatic carriers of pathogenic variants associated with frontotemporal degeneration. Neurology. 2024 Jan 23;102(2):e207926. doi: 10.1212/WNL.0000000000207926.

Pasrija P, Jha P, Upadhyaya P, Khan MS, Chopra M. Machine learning and artificial intelligence: A paradigm shift in big data-driven drug design and discovery. Curr Top Med Chem. 2022;22(20):1692–1727. doi: 10.2174/1568026622666220701091339.

Patel, L, Shukla, T, Huang, X, Ussery, DW, Wang, S. Machine learning methods in drug discovery. Molecules 2020;25:5277. doi: 10.3390/molecules25225277.

Peng C, Trojanowski JQ, Lee VM. Protein transmission in neurodegenerative disease. Nat. Rev. Neurol. 2020;16:199–212. doi: 10.1038/s41582-020-0333-7.

Prajjwal P, Flores Sanga HS, Acharya K, Tango T, John J, Rodriguez RSC, Dheyaa Marsool Marsool M, Sulaimanov M, Ahmed A, Hussin OA. Parkinson's disease updates: Addressing the pathophysiology, risk factors, genetics, diagnosis, along with the medical and surgical treatment. Ann Med Surg (Lond). 2023 Aug 8;85(10):4887–4902. doi: 10.1097/MS9.0000000000001142.

Pressman PS, Montembeault M, Matthewson G, Lemieux E, Brusilovsky J, Miller BL, Gorno-Tempini ML, Rankin K, Levenson RW. Conversational turn-taking in frontotemporal dementia and related disorders. J Neurol Neurosurg Psychiatry. 2024 Jan 11;95(2):197–198. doi: 10.1136/jnnp-2023-331389.

Querry M, Blanc F, Bousiges O, Philippi N, Cretin B, Demuynck C, Muller C, Botzung A. Memory outcome in prodromal and mild dementia with Lewy bodies and Alzheimer's disease: A longitudinal study. J Alzheimers Dis. 2023;94(1):147–162. doi: 10.3233/JAD-221243.

Ramezani M, Takian A, Bakhtiari A, Rabiee HR, Ghazanfari S, Mostafavi H. The application of artificial intelligence in health policy: A scoping review. BMC Health Serv Res. 2023 Dec 15;23(1):1416. doi: 10.1186/s12913-023-10462-2.

Rani N, Alam MM, Jamal A, Bin Ghaffar U, Parvez S. Caenorhabditis elegans: A transgenic model for studying age-associated neurodegenerative diseases. Ageing Res Rev. 2023 Nov;91:102036. doi: 10.1016/j.arr.2023.102036.

Renton AI, Dao TT, Johnstone T, Civier O, Sullivan RP, White DJ, Lyons P, Slade BM, Abbott DF, Amos TJ, Bollmann S, Botting A, Campbell MEJ, Chang J, Close TG, Dörig M, Eckstein K, Egan GF, Evas S, Flandin G, Garner KG, Garrido MI, Ghosh SS, Grignard M, Halchenko YO, Hannan AJ, Heinsfeld AS, Huber L, Hughes ME, Kaczmarzyk JR, Kasper L, Kuhlmann L, Lou K, Mantilla-Ramos YJ, Mattingley JB, Meier ML, Morris J, Narayanan A, Pestilli F, Puce A, Ribeiro FL, Rogasch NC, Rorden C, Schira MM, Shaw TB, Sowman PF, Spitz G, Stewart AW, Ye X, Zhu JD, Narayanan A, Bollmann S. Neurodesk: An accessible, flexible and portable data analysis environment for reproducible neuroimaging. Nat Methods. 2024 Jan 8;21(5):804–808. doi: 10.1038/s41592-023-02145-x.

Rossetti G, Kless A, Lai L, Outeiro TF, Carloni P. Investigating targets for neuropharmacological intervention by molecular dynamics simulations. Biochem Soc Trans. 2019 Jun 28;47(3):909–918. doi: 10.1042/BST20190048.

Ruffini N, Klingenberg S, Schweiger S, Gerber S. Common factors in neurodegeneration: A meta-study revealing shared patterns on a multi-omics scale. Cells. 2020 Dec 8;9(12):2642. doi: 10.3390/cells9122642.

Sarkar C, Das B, Rawat VS, Wahlang JB, Nongpiur A, Tiewsoh I, Lyngdoh NM, Das D, Bidarolli M, Sony HT. Artificial intelligence and machine learning technology driven modern drug discovery and development. Int J Mol Sci. 2023 Jan 19;24(3):2026. doi: 10.3390/ijms24032026.

Sasikumar S, Strafella AP. The challenging quest of neuroimaging: From clinical to molecular-based subtyping of Parkinson disease and atypical parkinsonisms. Handb Clin Neurol. 2023;192:231–258. doi: 10.1016/B978-0-323-85538-9.00004-3.

Schneider, P, Walters, WP, Plowright, et al. Rethinking drug design in the artificial intelligence era, Nat. Rev. Drug Discov. 2020, 19, 353–364. doi: 10.1038/s41573-019-0050-3.

Schwalbe N, Wahl B. Artificial intelligence and the future of global health. Lancet. 2020 May 16;395(10236):1579–1586. doi: 10.1016/S0140-6736(20)30226-9.

Shi Z, Jiang B, Liu T, Wang L, Pei G, Suo D, Zhang J, Funahashi S, Wu J, Yan T. Individual-level functional connectomes predict the motor symptoms of Parkinson's disease. Cereb Cortex. 2023 May 9;33(10):6282–6290. doi: 10.1093/cercor/bhac503.

Silva-Spínola A, Baldeiras I, Arrais JP, Santana I. The road to personalized medicine in Alzheimer's disease: The use of artificial intelligence. Biomedicines. 2022 Jan 29;10(2):315. doi: 10.3390/biomedicines10020315.

Son J, Kim D. Applying network link prediction in drug discovery: An overview of the literature. Expert Opin Drug Discov. 2024 Jan–Jun;19(1):43–56. doi: 10.1080/17460441.2023.2267020.

Soto C, Pritzkow S. Protein misfolding, aggregation, and conformational strains in neurodegenerative diseases. Nat. Neurosci. 2018;21:1332–1340. doi: 10.1038/s41593-018-0235-9.

Swalley SE. Expanding therapeutic opportunities for neurodegenerative diseases: A perspective on the important role of phenotypic screening. Bioorg Med Chem. 2020 Feb 1;28(3):115239. doi: 10.1016/j.bmc.2019.115239.

Talevi A. Computer-aided drug discovery and design: Recent advances and future prospects. Methods Mol Biol. 2024;2714:1–20. doi: 10.1007/978-1-0716-3441-7_1.

Teleanu DM, Niculescu AG, Lungu II, Radu CI, Vladâcenco O, Roza E, Costăchescu B, Grumezescu AM, Teleanu RI. An overview of oxidative stress, neuroinflammation, and neurodegenerative diseases. Int J Mol Sci. 2022 May 25;23(11):5938. doi: 10.3390/ijms23115938.

Teli P, Kale V, Vaidya A. Beyond animal models: Revolutionizing neurodegenerative disease modeling using 3D in vitro organoids, microfluidic chips, and bioprinting. Cell Tissue Res. 2023 Oct;394(1):75–91. doi: 10.1007/s00441-023-03821-2.

Tzeplaeff L, Wilfling S, Requardt MV, Herdick M. Current state and future directions in the therapy of ALS. Cells. 2023 May 31;12(11):1523. doi: 10.3390/cells12111523.

Ulugut H, Pijnenburg YAL. Frontotemporal dementia: Past, present, and future. Alzheimers Dement. 2023 Nov;19(11):5253–5263. doi: 10.1002/alz.13363.

Vamathevan J, Clark D, Czodrowski P, Dunham I, Ferran E, Lee G, Li B, Madabhushi A, Shah P, Spitzer M, Zhao S. Applications of machine learning in drug discovery and development. Nat Rev Drug Discov. 2019 Jun;18(6):463–477. doi: 10.1038/s41573-019-0024-5.

Vatansever S, Schlessinger A, Wacker D, Kaniskan HÜ, Jin J, Zhou MM, Zhang B. Artificial intelligence and machine learning-aided drug discovery in central nervous system diseases: State-of-the-arts and future directions. Med Res Rev. 2021 May;41(3):1427–1473. doi: 10.1002/med.21764.

Vemula D, Jayasurya P, Sushmitha V, Kumar YN, Bhandari V. CADD, AI and ML in drug discovery: A comprehensive review. Eur J Pharm Sci. 2023 Feb 1;181:106324. doi: 10.1016/j.ejps.2022.106324.

von der Recke F, Warmerdam E, Hansen C, Romijnders R, Maetzler W. Reduced range of gait speed: A Parkinson's disease-specific symptom? J Parkinsons Dis. 2023;13(2):197–202. doi: 10.3233/JPD-223535.

Wang H, Fu T, Du Y, Gao W, Huang K, Liu Z, Chandak P, Liu S, Van Katwyk P, Deac A, Anandkumar A, Bergen K, Gomes CP, Ho S, Kohli P, Lasenby J, Leskovec J, Liu TY, Manrai A, Marks D, Ramsundar B, Song L, Sun J, Tang J, Veličković P, Welling M, Zhang L, Coley CW, Bengio Y, Zitnik M. Scientific discovery in the age of artificial intelligence. Nature. 2023 Aug;620(7972):47–60. doi: 10.1038/s41586-023-06221-2.

Weeda MM, van Nederpelt DR, Twisk JWR, Brouwer I, Kuijer JPA, van Dam M, Hulst HE, Killestein J, Barkhof F, Vrenken H, Pouwels PJW. Multimodal MRI study on the relation between WM integrity and connected GM atrophy and its effect on disability in early multiple sclerosis. J Neurol. 2024 Jan;271(1):355–373. doi: 10.1007/s00415-023-11937-2.

Wu D, Chen Q, Chen X, Han F, Chen Z, Wang Y. The blood-brain barrier: Structure, regulation, and drug delivery. Signal Transduct Target Ther. 2023 May 25;8(1):217. doi: 10.1038/s41392-023-01481-w.

Yang X, Huang K, Yang D, Zhao W, Zhou X. Biomedical big data technologies, applications, and challenges for precision medicine: A review. Glob Chall. 2023 Nov 20;8(1):2300163. doi: 10.1002/gch2.202300163.

Yang, F, Zhang, Q, Ji, X, Zhang, Y, Li, W, Peng, S, Xue, F. Machine learning applications in drug repurposing. Interdiscip. Sci. 2022, 14, 15–21. doi: 10.1007/s12539-021-00487-8.

Zagórska A, Czopek A, Fryc M, Jaromin A, Boyd BJ. Drug discovery and development targeting dementia. Pharmaceuticals (Basel). 2023 Jan 19;16(2):151. doi: 10.3390/ph16020151.

Zhang L, Chen Y, Liang X, Wang L, Wang J, Tang Y, Zhu X. Prediction of quality of life in patients with Parkinson's disease with and without excessive daytime sleepiness: A longitudinal study. Front Aging Neurosci. 2022 Apr 13;14:846563. doi: 10.3389/fnagi.2022.846563.

Zhao Y, Ray A, Portengen L, Vermeulen R, Peters S. Metal exposure and risk of Parkinson disease: A systematic review and meta-analysis. Am J Epidemiol. 2023 Jul 7;192(7):1207–1223. doi: 10.1093/aje/kwad082.

7 Ethical Considerations in AI Implementation for Neurodegenerative Disease Diagnosis and Treatment

Madan Mohan Tito Ayyalasomayajula

7.1 INTRODUCTION

7.1.1 Why Is Neurodegenerative Disease Management an Area of Intrigue for AI Experts?

Neurodegenerative disease management poses significant challenges for healthcare practitioners owing to diseases' gradual progression and rising occurrence in the elderly population. These disorders, including Alzheimer's disease, Parkinson's disease (PD), and Huntington's disease, are characterized by the degeneration and loss of neurons in the brain, resulting in cognitive deficits, movement dysfunctions, and other incapacitating symptoms. While various therapies may provide respite for specific physical or mental symptoms, blocking their advancement is still challenging, and no definitive treatments are available. The likelihood of developing a neurodegenerative disease considerably rises with advancing age, which also contributes to the decreased lifespan of the affected population. Artificial intelligence (AI) and machine learning (ML) technologies have gained significant attention in neurodegenerative disease research to formulate new strategies to address these challenges. Clinically, these disorders present diverse characteristics, indicating that they impact people differently depending on age, gender, and heredity. Designing individualized treatment plans that address each patient's unique requirements is critical to maximize results (Stefano, 2023).

Neurodegeneration requires advanced computational analysis to understand its root causes and stage of progression and develop accurate diagnostic and treatment approaches. Researchers have employed ML methods, profound learning models like neural networks, and sophisticated coupled architectures in combination with neuroimaging and motion-tracking technology. It has been demonstrated that these algorithms can analyze symptoms and biomarkers that are pertinent to pathology, track changes in the brain's anatomy and function that are related to disorders, and track patients' motor and cognitive signs in conjunction with how they respond to intervention (Khaliq et al., 2023). They provide prospective pathways for managing massive datasets derived from various sources, including genetic sequencing, electronic health records, and medical imaging modalities.

Further, AI can assess patient data and recommend specific approaches that contribute to precision medicine. Deep learning (DL) models created for analyzing magnetic resonance imaging (MRIs) allow early identification of Alzheimer's disease, which aids in disease management. In addition, this emerging technology contributes to drug research and development, assisting in enhancing neurodegenerative therapeutics. Additionally, researchers employ natural

DOI: 10.1201/9781032711102-7

language processing (NLP) algorithms to extract valuable information from speech patterns to assess the extent of non-motor impairment in PD (Pérez-Toro et al., 2019). Integrating AI into treating neurodegenerative diseases has many benefits, including earlier and more accurate diagnosis, improved patient outcomes, and decreased healthcare costs. As AI advances, it will increasingly play a more prominent role in addressing neurodegenerative disorders, providing renewed optimism for those impacted by these incurable ailments (Chatterjee, 2021).

7.1.2 Overview of Ethical Issues in AI Application

AI and neurotechnology showcase great potential for enhancing our understanding and the treatment of neurodegenerative disorders. Nevertheless, incorporating these technologies gives rise to noteworthy ethical concerns, namely, with the protection of data privacy, the autonomy of patients, the potential bias of algorithms, the extent of augmentation, and the level of transparency. The aggregation of brain data can reveal personal and private information about individuals, highlighting the need for strong safeguards against illegal access and improper usage. With the advancement of neurotechnology, traditional ideas of identity and control could be questioned, which means there is a need for specific standards to protect fundamental human rights. Furthermore, if not subjected to strict regulation, augmentation has the potential to exacerbate social inequalities and perpetuate discriminatory practices. Prejudiced data structures and obscure algorithms pose threats to impartiality and equality. To tackle these ethical concerns, it is essential to have ongoing discussions that involve politicians, ethicists, scientists, and the affected communities. By cultivating transparent and all-encompassing dialogues and adopting optimal methodologies, we can establish confidence, advocate for fair opportunities, and reduce negative consequences (Yuste et al., 2017).

7.2 ETHICAL DILEMMAS

7.2.1 Ethical Dilemmas Surrounding AI-Driven Diagnoses

AI, ML, and DL models have shown promise in assessing various data sources related to neurodegenerative disorders, such as genetic information, neuroimaging scans, and clinical evaluations. These algorithms use extensive datasets to detect subtle patterns and biomarkers, facilitating earlier and more accurate diagnosis. Moreover, automated diagnostic technologies can speed up the diagnosis process and reduce the workload of healthcare workers. ML techniques, such as Naive Bayes, decision tree, support vector machine (SVM), random forest (RF), K-nearest neighbors (KNN), and ensemble models, have proven to be effective in different areas of clinical diagnosis, such as gait analysis, genetic identification, neuroimaging classification, and disease progression prediction. AutoML enhances diagnostic tools by automating, selecting, and optimizing models, which might allow for remote evaluation of these neurological conditions (Ganesh et al., 2023). Privacy issues related to patient information contribute to data isolation since strict restrictions and ethical concerns discourage the unrestricted sharing of genetic and clinical data. This situation creates a substantial obstacle to gaining comprehensive knowledge and implementing appropriate treatment strategies for these disorders (Ganesh et al., 2023).

Another ethical concern is to provide openness and explainability in the algorithms used for diagnosis and prognosis since opaque or "black box" models may cause distrust and misunderstanding among patients and healthcare practitioners. Furthermore, there is a rising dispute about the role of people versus computers in healthcare, with some claiming that

AI and ML might completely replace physicians, raising concerns about patient autonomy and the doctor-patient relationship. There are also issues regarding algorithmic bias, which may lead to unfair or incorrect diagnoses based on criteria like age, color, or socioeconomic background. As AI, DL, and ML continue to change the future of healthcare, legislators, ethicists, and industry experts must work together to address ethical issues and develop explicit rules for the responsible use of these technologies.

For instance, Google Brain's DeepMind team developed AlphaFold, an AI model that employs DL to predict protein shapes (Jumper et al., 2021). This finding offers the potential to accelerate drug discovery and development for a wide range of disorders, such as those that result from defective protein folding, such as Alzheimer's and Parkinson's. Similar studies conducted at Massachusetts Institute of Technology (MIT) and Harvard University used ML and functional magnetic resonance imaging (fMRI) (Chatterjee et al., 2018) to predict the development of neurodegeneration up to six years before symptoms appeared. Such discoveries have enormous potential to enhance patient outcomes and our understanding of neurodegenerative disorders. For an appropriate and fair deployment of these technologies, it is critical to recognize and address the ethical issues concerning data sharing, transparency, and bias (Chudzik et al., 2024).

7.2.2 Ethical Dilemmas Surrounding AI-Driven Therapeutic Methodologies

Technological innovations are causing treatment approaches for neurological diseases to evolve more rapidly than ever before. One such instance is deep brain stimulation (DBS), a therapeutic method that employs ML algorithms to improve stimulus settings for each unique patient. DBS has been found to alleviate motor symptoms in PD and minimize medication usage (Machado & Reppold, 2015). Another field where AI is being investigated is drug development, with businesses such as Insilico Medicine using generative adversarial networks (GANs) to create tiny compounds with desirable qualities for target proteins linked with neurodegenerative disorders (Haider et al., 2021).

Devices that utilize brain-computer interface (BCI) technology enable direct contact between the brain and external equipment, creating prospects for regaining lost capabilities and developing new treatment procedures (Hummadi & Chatterjee, 2021). In stroke rehabilitation, BCI systems like the robot-assisted rehabilitation system (RARS) use AI methods such as neural networks and reinforcement learning to help patients recover. RARS, which combines BCI with robotic limb movement, aids in motor function restoration by analyzing brain signals collected with electroencephalography (EEG) caps. The device gives continuous input and tailors workouts to individual requirements, resulting in better results than traditional treatment. Similarly, BCI technology is used in DBS for PD treatment, which involves implanting electrodes to control aberrant brain activity. ML methods like SVMs and reinforcement learning improve therapeutic effectiveness by optimizing stimulation settings. Despite its invasive nature, DBS successfully treats symptoms such as tremors and stiffness. Ongoing research aims to use AI algorithms to improve treatment results (Machado & Reppold, 2015).

While these technologies have enormous promise for improving neurodegenerative disease therapies, they also present ethical problems. Confidentiality issues arise when dealing with critical patient data, especially regarding neuro data exchange and commercial exploitation. An additional issue is the absence of openness and comprehensibility around AI algorithms, making it difficult for healthcare practitioners and patients to grasp the logic behind these systems' judgments. Furthermore, constructing "designer brains" by focused neuromodulation poses concerns about identity and autonomy.

7.3 ETHICAL CONCERNS ARISING FROM BIASES IN AI MODELS

When the training data does not adequately reflect the population that is being targeted, AI algorithms are prone to algorithmic bias. In addition to gender, color, socioeconomic position, and other characteristics included in the data, this bias may exhibit itself in various ways. Algorithmic bias can be caused by inaccuracies in data collecting, which might take the form of human interpretations or historical preconceptions. Various demographic groups will likely have varying degrees of success in generalizing the diagnostic accuracy of prediction models. There is a growing concern about the accidental introduction of unfairness, which may lead to discrimination against underserved and underprivileged populations. ML algorithms can potentially improve clinical decision-making and public healthcare (Gichoya et al., 2023). Regarding decision-making processes, equity involves preventing bias or favors determined by attributes that are either innate or the result of learning. To fully comprehend the many manifestations of algorithmic bias, it is essential to acknowledge its difficulties, as described below (Figure 7.1).

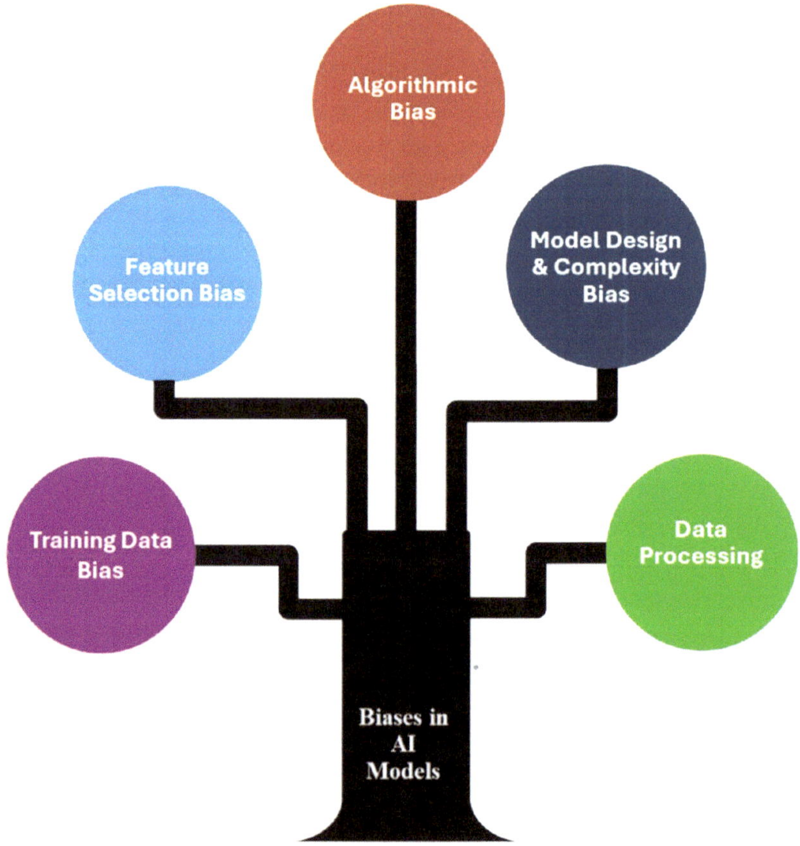

FIGURE 7.1 Biases in AI models.

(From European Commission. (2019). ***Ethics guidelines for trustworthy AI.*** https://digital-strategy.ec.europa.eu/en/library/ethics-guidelines-trustworthy-ai)

7.3.1 Training Data Bias

A significant limitation of AI in the context of neurodegenerative illnesses is the presence of bias in the training data. This limitation indicates that the datasets used for training AI models are inadequately balanced and do not precisely reflect the demographics of the real-world population. Biases of this kind may result in severe consequences, such as incorrect diagnosis and poor medical interventions, worsening the existing inequalities. Flaws in training data could emerge as gender bias if the training data for Alzheimer's disease was primarily based on male cases. Regarding biomarkers, the image recognition technique is likely to overlook symptoms that are more common in women, such as increased susceptibility to anxiety and depression rather than memory problems. This omission might be potentially harmful, considering that women have a greater susceptibility to Alzheimer's disease compared to males. Missing populations may also be an issue, like in the case of a dataset that broadly includes a specific race or age group of individuals. In such instances, the image recognition tool may have difficulty accurately diagnosing symptoms, particularly among those distinctive populations (Celi et al., 2022). For example, suppose the tool is used on a group of patients who are primarily Caucasian. In that case, the algorithm will probably be unable to identify slight differences in brain scans that are more common in the Asian or African American population.

7.3.2 Future Selection Bias

Including sensitive/discriminatory features in healthcare AI models can lead to biased outcomes. Predicting these outcomes would, in turn, result in the reinforcement of existing healthcare disparities. For example, if the model uses zip code as one of the features, the AI model's outcomes would worsen disparities in access to earlier described care regions that affect the health of populations residing in those locations differently. In reality, such a prediction could not be ethical and represents a failure to achieve non-discrimination and fairness in healthcare. Irrelevant features included in the input data can also corrupt an AI model. For instance, in the training model, selecting first name, last name, and profession is impertinent, which can result in a higher percentage of false negatives (Celi et al., 2022).

7.3.3 Algorithmic Bias

Errors in ML models due to biased data can occur when the AI models keep repeating systematic errors within them. For example, if a patient called "Bob" has a neurodegenerative disorder in the training dataset, this might lead to biased predictions by the model. In case the model wrongly learns that this means that any other person with these family names has the same condition, it classifies them as such even though their medical information does not suggest so. This kind of bias results in false-positive diagnoses where individuals are falsely identified as having diseases based on their non-pathological or disease-unrelated characteristics, including family relations. Such biases undermine AI algorithms' fairness and reliability, leading to wrong diagnoses and inappropriate treatment procedures. Avoiding algorithmic bias requires deliberations about what is included in the dataset, how features are selected, and which techniques are used for modeling because medical information must be given priority over coincidental relationships while making predictions about patients' conditions (Williams, 2023).

Due to biased data, ML models may yield erroneous outcomes if the AI model keeps repeating systematic errors within them. For instance, the model is biased if one patient with the surname "Bob" has a neurodegenerative disorder and the last name is used in the training dataset. Suppose the model assigns higher weightage to the last name and includes this as part of its learned knowledge. In that case, it can yield incorrect outcomes even though the information in the medical records does not support it for other patients with similar last names. This improper learning can lead to incorrectly classifying individuals using AI and ML algorithms and result in a false-positive scenario.

7.3.4 Model Design and Complexity Bias

The structure and intricacy of AI models play a crucial role in developing customized therapeutic approaches for neurodegenerative disorders. Insufficient oversight and verification of AI models with complex structures might magnify the influence of data bias, possibly leading to unfair treatment of patients. Take, for instance, a prediction model specifically created to evaluate the likelihood of certain diseases, which can display partiality toward certain factors. This bias may arise from the model's restricted comprehension of the many expressions of neurological disorders, resulting in an unequal impact on specific demographic groups and ethical considerations related to the principles of doing good and avoiding harm (Gichoya et al., 2023). These biases worsen the risks associated with pharmaceutical treatment options, unfairly amplifying them. Hence, it is crucial to meticulously evaluate the structure and complexity of AI models before using them to manage neurodegenerative disorders. This evaluation involves refraining from using models that are too simplified and implementing comprehensive validation methods to minimize bias successfully.

7.3.5 Data Preprocessing

Extensive exploration of AI algorithm biases is crucial for fair and effective healthcare. Biased assumptions in data preparation approaches like imputation and normalization might worsen health inequities, especially for underrepresented populations. It may influence health results, causing injury and posing ethical questions about nonmaleficence. Biased AI systems perpetuate healthcare inequality beyond inaccurate predictions. AI medical diagnostic systems' repeated misdiagnoses might lead to inappropriate treatments, violating autonomy and fairness. Distorted algorithms hinder informed choice and discriminate against disadvantaged people, undermining the right to healthcare. Recognizing these challenges necessitates proactive initiatives to eliminate AI prediction biases, promote democratic healthcare, and respect medical ethics. Upholding justice, beneficence, and respect for humans requires a commitment to removing systematic flaws in AI algorithms to provide trustworthy and impartial evaluations (Gichoya et al., 2023). Researchers and practitioners may collaborate to create transparent algorithmic defect detection and correction systems that improve medical practice while honoring ethics.

7.4 DATA OWNERSHIP, PRIVACY, AND SECURITY

Neurodegenerative diseases such as Alzheimer's, Parkinson's, and amyotrophic lateral sclerosis present distinct issues in terms of data possession, security, and security.

Researchers depend largely on vast datasets of electroencephalogram (EEG), MRI, and other forms of data to create AI models capable of forecasting disease progression, detecting the presence of biomarkers, and devising successful therapeutic methodologies. However, data acquisition, data management, and data protection of these databases present several ethical considerations (Baines et al., 2024). In this context, we will look mainly at the ownership and personally identifiable information (PII) issues surrounding EEG data, as well as potential solutions and recommended procedures for managing this private data ethically.

7.5 CHALLENGES WITH LARGE DATASETS FOR AI MODEL DEVELOPMENT

Using large datasets to build AI models for neurodegenerative diseases allows these systems to analyze extensive category data, improving dependability and accuracy. However, despite these benefits, large databases raise ethical questions about data ownership, confidentiality, and safety. Since AI models use several databases, appropriate access to sensitive medical data is crucial. Due to conflicting claims between medical practitioners, research entities, and other stakeholders, determining possession of EEG arrays from patients or trial participants is difficult. Addressing these moral barriers assures responsible and equitable AI implementation in healthcare.

Protecting the PII inherent in these datasets is critical for maintaining patients' privacy rights and fostering confidence among all parties. Ensuring robust data handling protocols and adhering to legal frameworks such as the Health Insurance Portability and Accountability Act (HIPAA) are critical steps toward reducing the risks associated with unauthorized access and mistreatment of sensitive information. Recognizing the ethical challenges arising from the requirement for big, varied datasets to feed AI model development allows researchers and professionals to collaborate to establish data acquisition and management standards. This standardization involves transparent discourse, informed approval, and explicit data use and sharing procedures. Ultimately, these initiatives will help to ensure the proper use of AI technology in neurodegenerative disease diagnosis, treatment, and research (Munappy et al., 2019).

7.6 MEASURES TO MITIGATE ETHICAL DILEMMAS

7.6.1 Informed Consent

When considering the ethical and legal challenges associated with managing sensitive information by healthcare institutions, researchers, and administrations, it is impossible to understate the urgent demand for upgrading informed consent for using AI technology in healthcare. Particularly in light of recent events, such as the controversy surrounding Cambridge Analytica (Isaak & Hanna, 2018) and Google's purchase of the personal information of minors without the appropriate authorization (Natasha & Conger, 2019), there has been a growing desire for more transparent and verifiable approval about the use of data. Even though stricter restrictions could provide some relief, the fundamental problem still exists: the need to preserve human liberty and comprehension via the implementation of informed consent processes that are carried out effectively.

Effective communication, knowledge, and acceptance of one's decision-making ability are vital elements of informed consent in the healthcare industry. However, standard practices in big data, essential for developing AI models, often neglect to get informed consent, resulting in privacy violations and unapproved use of data. Stakeholders are pushing for the ethical integration of AI as it is growing in popularity in the healthcare industry, particularly in clinical research settings. This entails implementing user education campaigns and creating visually appealing legal paperwork. By ensuring patients' rights are respected and their privacy is maintained, these steps seek to increase patient trust when AI is incorporated into various healthcare professions.

According to Andreotta et al. (2022), the conventional one-time, static informed consent method must be updated to successfully address the ever-changing environment of AI and significant data consumption in healthcare. To close this gap, the authors suggest a more flexible and continuing method known as "Continuous consent" with characteristics that include,

a. Persistent Approval: This allows people to provide permission for their data to be used for specific purposes and to withdraw their consent at any moment.
b. Fine-tuned Access: This option allows users to create complex data access settings and control their use.
c. Live updates convey customers' real-time visibility into how their data is used in AI applications.
d. Ethics and Supervision: Integrates ethical norms and human interaction to enable responsible AI usage while protecting individual privacy.

Continuous consent addresses the challenges presented by the dynamic nature of AI technologies and a large amount of healthcare-generated data. Its benefits include increased openness, strong protection of rights and anonymity, and tighter connection with the growing fields of AI and big data. Despite facing legal and practical issues, the authors encourage academics and policymakers to collaborate to overcome these obstacles and implement continuous consent (Andreotta et al., 2021).

7.6.2 Data Security and Confidentiality

While AI has the potential to improve healthcare, substantial concerns remain. The challenges, ranging from questions about the integrity of medical data to privacy concerns and biases resulting from human inputs, need coherent and collaborative solutions.

Therefore, it is vital to cultivate cooperation across several disciplines, establish rigorous data annotation procedures, build efficient data transmission techniques, and progress algorithms within ML research and actual clinical applications. Computer technologists need to collaborate closely with medical practitioners to achieve the goal of effectively integrating AI into clinical settings. In addition, continuous training in AI is essential to ensure adaptability, compliance, and the avoidance of unanticipated legal concerns. Furthermore, it is needed to facilitate uninterrupted cooperation and data flow across various healthcare settings to fully capitalize on the promise of emerging technology to improve patient care and incorporate AI issues into current medical education courses.

Given the frequency and financial consequences of unlawful data breaches resulting in unauthorized access, viewing, theft, or private information, it is vital to base cybersecurity strategy decisions and operations on predictive analytics tools and preventative strategies. Such preventative actions may help reduce the potential harm caused by these illegal acts (Alfalahi et al., 2023). An essential accompaniment to developing and applying sophisticated AI algorithms in the healthcare industry is the conduct of public campaigns that advocate for awareness, transparency, and personal accountability.

These programs ensure that sensitive patient data is handled securely and ethically. Simultaneously, strategic collaborations across many sectors, including academics, government, and business, would reinforce security measures and allow smooth cooperation toward common objectives. Responsibly utilizing cutting-edge AI technology may result in significant advantages, including increased patient involvement, improved operational flow, more excellent research capabilities, and increased efficiency across the whole healthcare spectrum.

7.6.3 Transparent Data Sharing

Transparent data sharing policies can ensure trust and confidence in implementing AI in diagnosing and treating neurodegenerative diseases by providing data security and privacy. Such policies may include how data is collected, processed, shared, and secured in the AI environment. The Europe General Data Protection Regulation has policies establishing explicit consent requirements, limiting data usage for explicit purposes, and data on security and storage mechanisms. The regulation also allows data portability, implying that individuals can request their data from an organization or people who transfer it. However, the United States may rely on the HIPAA, which protects and confidentially handles individually identifiable information (Hlávka, 2020). The act calls for written authorization to disclose personal health information (PHI), which takes place to authorized personnel who access the record on a need-to-know basis and ensures that the EHR data is password protected. The European Open Science Cloud and the National Institutes of Health's Big Data to Knowledge have also utilized technology to promote transparent data-sharing initiatives in scientific research for neurodegenerative diseases. This enables researchers to store, manage, analyze, and publish the data without compromising privacy or secure data. Therefore, all AI professionals who diagnose or treat neurodegenerative diseases should respect and adhere to the policies.

7.6.4 Transparency and Interpretability in AI Systems: With Explainable AI Approaches

AI has made significant improvements in a range of fields, including healthcare, where it has the potential to revolutionize diagnostics and therapy planning. However, the growing use of AI in healthcare poses ethical concerns about transparency, fairness, and responsibility. Explainable AI (XAI) is a branch of AI research that focuses on creating intelligent systems whose decision-making processes can be easily understood, grasped, and trusted by experts and patients (Mersha et al., 2024). These AI systems can deliver clear, concise, and intelligible explanations for their actions, recommendations, and projections. These explanations assist clinicians in understanding the underlying rationale and context of the AI's decisions, boosting confidence, reducing ambiguity, and increasing overall trustworthiness.

Regarding ethical healthcare AI, XAI is vital in addressing various significant concerns.

Patient Safety: Understanding how AI systems make judgments is critical to ensure they operate safely and effectively. XAI enables doctors and other healthcare professionals to evaluate AI systems' thinking processes, identify potential risks, and take appropriate action. For example, clarifying the reasons behind a diagnostic recommendation may help professionals validate the accuracy of the finding and prevent misdiagnosis.

Fairness and Equality: Building AI systems requires treating all individuals equally and impartially, regardless of demographic differences or sensitive features. XAI ensures AI decisions are free of hidden biases and prejudice by identifying the factors that impact such results. By providing detailed explanations, healthcare practitioners may examine the fairness and equality of AI recommendations and make changes as appropriate.

Privacy Protection: Keeping patients anonymous is crucial in healthcare. XAI makes it simpler to create privacy-preserving AI systems by providing fine-grained control over the data required to make choices. Healthcare organizations may preserve patients' privacy while offering high-quality care by stating which data pieces lead to a specific proposal. Clinical research and pharmaceutical development may be significantly expedited using XAI, which helps researchers assess the effect of different therapies and medications on patient populations. By analyzing enormous amounts of information and creating explanatory reports, XAI may reveal patterns, trends, and connections that might otherwise go undiscovered, leading to faster discoveries and better cures.

The most well-known XAI strategies are rule-based methods, local interpretable model-agnostic explanations (LIME), DeepLIFT, and SHapley Additive exPlanations (SHAP). Each technique has unique features such as transparency, interpretability, and processing efficiency, making it a valuable tool for creating ethical healthcare AI systems. In a word, XAI is a critical step toward establishing trustworthy, transparent, and ethical AI systems for healthcare. XAI improves medical professionals' judgment, promotes justice and equality, protects data confidentiality, and speeds up AI diagnostic and therapeutic research by bridging the gap between technology and human understanding (Rasheed et al., 2022).

7.6.5 Human Oversight and Error Correction

Human supervision in AI-driven healthcare systems is crucial to ensure that AI technology is used appropriately and ethically. Healthcare professionals, legislators, and regulatory agencies must actively monitor and supervise the functioning of AI algorithms and systems. Here are some essential qualities of human supervision in AI-powered healthcare systems.

Accountability and Responsibility: Healthcare professionals play a vital role in overseeing AI algorithms and systems to ensure accountability for the decisions and acts made by AI. They are responsible for the outcomes and must be able to act and correct any errors or biases that arise. Human supervision promises that AI adheres to established medical norms, ethical standards, and legal limits. Human supervision is essential to examine and address ethical problems in AI-powered healthcare. It requires ensuring that AI algorithms and systems respect patient privacy, maintain confidentiality, and adhere to informed consent requirements. Human monitoring helps to identify and remove biases, prejudices, and inequities in AI systems, ensuring consistent and fair healthcare results.

Interpretability and Transparency: AI algorithms may act as "black boxes," making it untraceable to understand the reasoning behind their decisions. Human monitoring increases the transparency and interpretability of AI-driven decisions. Healthcare practitioners play a vital role in assessing and validating AI system outputs, confirming their consistency with clinical expertise, and explaining them to patients as appropriate.

AI algorithms are susceptible to errors, biases, and misinterpretations. Human supervision facilitates the identification and correction of these issues. Healthcare practitioners may investigate when AI-generated results contradict their clinical expertise, enabling them to intervene, correct, or retrain the AI algorithms as needed. Human monitoring helps to ensure that AI-powered healthcare solutions are accurate, safe, and trustworthy.

Contextual Knowledge and Skills: While AI algorithms are trained on data, they may lack the contextual knowledge and skills that healthcare practitioners possess. Human supervision provides the clinical judgment, intuition, and skill required to augment AI systems' capabilities. Healthcare staff may contextualize AI-generated insights within the background of overall patient care, considering individual patient characteristics, preferences, and unique scenarios.

Continuous Evaluation and Verification: Human supervision requires continual monitoring and validation of AI algorithms and systems. This supervision includes monitoring their performance, identifying any drift or loss in accuracy over time, and ensuring that the algorithms remain current on medical knowledge and practices. Human supervision helps to identify and resolve any errors, limitations, or biases that may exist in AI-powered healthcare systems.

In brief, human participation in AI-powered healthcare systems promotes openness, addresses ethical problems, increases interpretability and transparency, identifies and corrects mistakes, offers context-dependent expertise and knowledge, and enables continuous review and assurance. By combining the advantages of AI algorithms with human intelligence and understanding, we may achieve the optimum balance of current technology and human interaction, resulting in enhanced safety, efficiency, and patient-first practices in healthcare.

7.6.6 Training and Educating Users on AI Technology

Healthcare personnel need to get training and education to fully understand the use of ethical AI in healthcare and have confidence in making this implementation. The administration of healthcare is being revolutionized by AI, and such courses are vital for educating professionals to handle difficulties associated with responsibly integrating AI, commonly referred to as responsible AI. They provide the information, abilities, and self-assurance that are required. Companies in the healthcare industry have the potential to foster an environment of trust and collaboration between humans and machines by providing comprehensive training programs that include AI theory, ethical concerns, and practical applications. Not only does the deployment of AI technology enhance patient care by facilitating more effective decision-making and workflow efficiency, but it also guarantees an ethical and responsible implementation.

To advance the inclusion of AI in healthcare, acquiring information without interruption and collaborating across a wide range of academic disciplines is essential. In addition to initial training, by participating in ongoing education programs and collaborative platforms, healthcare professionals can maintain their knowledge of the most current developments and best practices in the AI domain. Businesses in the healthcare industry may further improve the capacities of professionals by investing in these initiatives. These investments will allow professionals to use AI fully, ultimately leading to improved patient outcomes, resolving ethical difficulties, and promoting innovation in healthcare delivery. It is possible for the healthcare community to effectively employ AI to satisfy the constantly evolving needs of patients and healthcare systems if we collaborate and commit ourselves to lifelong learning. These developments will pave the way for a future in which technological advancements will enhance the quality, accessibility, and efficiency of healthcare services (Table 7.1).

TABLE 7.1
Ethical Considerations: Strategies, Benefits, and Risks

Ethical Concern	Mitigation Strategy	Potential Benefit	Potential Downside
Bias in AI algorithms	• Use various training datasets that are representative of the population. • Use fairness measures in model development to identify and correct the bias. • Use human-in-the-loop decisions for crucial decisions on patient administration.	Improved diagnostic accuracy and early detection across the larger population.	Delays in treatment due to the requirement of human supervision.
Privacy and data security	• Obtain informed consent from patients who have reached the legal age of consent. • Obtain assent from parents or legal guardians when the patient cannot provide consent. • Anonymization or de-identification of patients' sensitive data. • Implementing robust processes, auditing, and cybersecurity measures to protect patients' sensitive and non-sensitive data.	Ability to develop sophisticated personalized treatment plans and effective therapies.	Potential data breaches due to cyber attacks could compromise patient privacy and lead to the generation of misleading AI models.
Algorithmic transparency and explainability	• Developing interpretable AI models that can explain the reasoning behind the decisions of the model. • Clarify to patients how AI was used to treat them; in other words, provide clear explanations to patients in layman's language on how AI technology is used in their treatment. • Creating an open channel of communication between clinicians, researchers, and patients on how AI algorithms work and how they assist in the treatment of various ailments.	Increased trust in AI-driven treatment decisions and improved patient engagement.	Inability to explain to the patient in a non-technical way how the complex model works.

(Continued)

TABLE 7.1 *(Continued)*
Ethical Considerations: Strategies, Benefits, and Risks

Ethical Concern	Mitigation Strategy	Potential Benefit	Potential Downside
Access and equity	• Ensuring the accessibility and affordability of AI-aided treatment to all patients. • Eliminating geographical disparities in accessibility to AI-empowered treatments. • Establish operational guidelines for reasonable or uniform allocation of AI resources.	A better reduction of healthcare disparities emerges from improved quality of life in all patients.	Challenges would include instituting and initiating means of enforcing the fair and just guidelines.
Excessive reliance on AI	• Respecting, valuing, and constantly seeking human expertise in diagnosis, treatment planning, and patient care. • Treating AI as a virtual assistant tool that complements clinicians' decision-making processes and not a replacement. • Constant monitoring and frequent reevaluation of AI models.	Higher efficiency, accuracy, and precision of clinical decision-making.	There are possible errors if AI models are improperly monitored and obsolete models are used.

7.7 REGULATORY COMPLIANCE AND LEGAL FRAMEWORKS

Regulatory compliance is a streamlined process that ensures adherence to laws, regulations, standards, and guidelines established by regulatory bodies or any legal entity backed by federal or governing bodies. These regulatory bodies ensure organizational practices are consistent with specific requirements through routine surveillance and auditing. Noncompliance with the set regulations may incur penalties or legal action.

In contrast, a legal framework refers to the entire system of laws, rules, statutes, and principles regulating a jurisdiction and providing guiding principles to operate on various social activities. It sets forth rights, obligations, and means of enforcement. It is the basis for regulatory requirements. Regulatory compliance zooms in on one regulation at a time, while the legal framework is the broad context for all these.

Governments commonly establish legal frameworks and regulatory compliance requirements to ensure transparency. Regulatory compliance is generally a subset of the broad spectrum of legal frameworks. Though nations set individual arrangements, discussing the widely known frameworks is noteworthy.

7.7.1 General Data Protection Regulation

The European Union (EU) established a critical regulatory and legal framework that protects the security and confidentiality of EU citizens' data, namely General Data Protection

Regulation (GDPR). The establishment was initiated in 2016 and was enacted in 2018 across the EU. GDPR sets the guiding and working principle for the organizations that collect personal data of EU citizens, and all the organizations that operate and process personal datasets within or out of the EU fall into the scope of this regulation, which, in other words, is effectively location-neutral (European Parliament and of the Council, 2016).

7.7.2 California Consumer Privacy Act

The California Consumer Privacy Act (CCPA) is a detailed privacy law attributed to California's residents. The CCPA was established in 2018 and was enacted by policymakers in 2020 (State of California Department of Justice Office of the Attorney General, 2023). The CCPA enables individuals with specific rights regarding personal data and forces organizations to enforce compliance factors on the collected data points. This law is the first of its kind in the data privacy space in the United States. California introduced it, and other states are following in its footsteps. CCPA is very similar to the GDPR, supporting the right to access, right to correct, right to delete, right to opt-out, right to limit the use of sensitive information, etc. (Leginfo.legislature.ca.gov, 2018).

7.7.3 Role of CCPA And GDPR in AI Models

Key points to be addressed as part of GDPR and CCPA are ensuring the rights of data subjects, following data protection principles, maintaining accountability and compliance, and notifying regulatory authorities of data breaches. The Rights of Data Subjects permit EU citizens or individuals to access, update, or delete their data processed by firms. In a scenario where individuals are adding, updating, or deleting their personal information where the data is used in the AI models to train for either personalization or any recommendation or to serve any organizational use cases, then the firms are bound to retrain their AI models as the data turns to become obsolete and raises the ethical concerns. This retraining process is further complicated when an individual requests the "Right to Rectify" or "Right to Delete."

Regarding the data protection principles governed by the GDPR and CCPA, organizations must process individuals' personal information for a legitimate need by following transparency, fairness, accuracy, integrity, confidentiality, purpose limitation, and data minimization principles. AI processes often demand more data features and extensive data volumes to be more effective and accurate in learning, improving their accuracy, and providing predictions or classifications. Overlooking may conflict with the Data Minimization principle set in GDPR and CCPA, and over-utilizing the datasets in AI models may lead to raising red flags for ethical and compliance issues.

It is essential to note that neither the GDPR nor the CCPA explicitly expresses specific guidelines for using AI models. While businesses process individuals' data, all the regulations applicable to computational systems are equally applicable even when implementing AI models.

7.7.4 Ethical Guidelines for Trustworthy AI by the European Commission

The Ethical Guidelines for Trustworthy AI, established by the European Commission's High-Level Expert Group on Artificial Intelligence (AI HLEG) in June of 2018, laid the foundation for developing AI technologies responsibly within the EU. While non-binding,

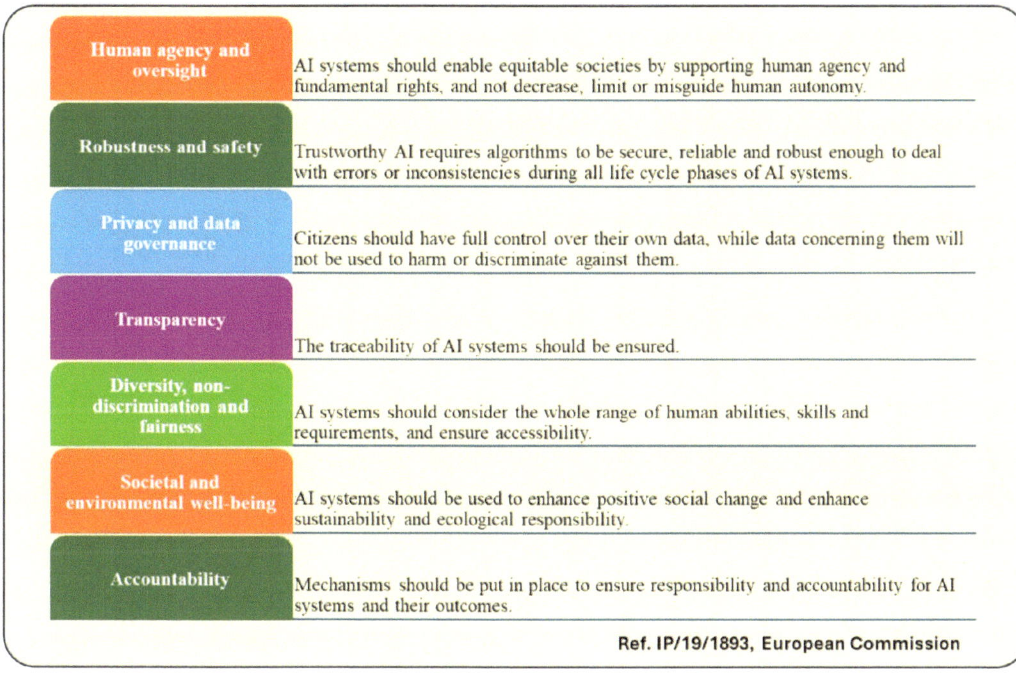

FIGURE 7.2 Seven essentials for achieving trustworthy AI. (From European Commission. (2019). *Artificial intelligence: Commission takes forward its work on ethics guidelines.* https://ec.europa.eu/commission/presscorner/detail/en/IP_19_1893)

the guidelines published the following year have played a crucial role in policymaking and industry practices relating to ethics in AI. The comprehensive framework addresses transparency, diversity, human oversight, privacy, and social welfare concerns to promote accountable and fair design from the start (European Commission, 2019).

While previous approaches often focused narrowly on technical performance, AI HLEG emphasized a multidisciplinary vision for AI that considers legal and societal impacts with the same care as results or functions. Their work has raised essential discussions around ensuring AI benefits rather than harming humanity or undermining fundamental rights and democratic values. As adoption continues staggeringly, these voluntary principles provide orientation for anyone seeking to understand and address ethical dilemmas arising from our increasingly intelligent creations. Below are the seven critical essentials set as the guiding principles for achieving trustworthy AI (Figure 7.2).

7.8 CASE STUDY 1: ETHICAL ISSUES AND SOLUTIONS IN PARKINSON'S DISEASE MANAGEMENT WITH WEARABLE DEVICES

PD affects millions of people globally, putting enormous pressure on healthcare systems. While a definitive cure remains unclear, effectively managing symptoms is critical to improving patients' quality of life. Wearable gadgets have emerged as a possible route for PD care, providing continuous monitoring, symptom tracking, and potential treatment interventions. However, this integration creates various ethical considerations that must be carefully investigated and addressed.

A comprehensive study was conducted by Kenny et al. (2022) to investigate the perspectives and requirements of individuals who suffer from PD in relation to the use of AI-powered wearable devices for disease monitoring. Their research findings shed light on the critical importance of patient-centered design and the need to maintain open lines of communication between patients and developers throughout the design and development process. Within their encounters with healthcare providers, patients have shown a significant need for privacy, autonomy, comfort, and user-friendly interfaces. Incorporating these fundamental aspects is crucial to developing a sense of ownership and control over managing their health. On the other hand, patients advocated for clinical accuracy in wearable devices, underlining the need to boost professional evaluations rather than replacing them via wearable technology.

7.8.1 Patient Needs and Priorities

Featured gadgets allow individuals to manage their health with ownership and control. Furthermore, clinical precision is crucial. Patients desire wearables that deliver accurate data that supplements, rather than replacing, their professional evaluations. Ideally, these gadgets would eliminate the need for repeated hospitalizations, which are a severe hardship for many patients with PD. Treating PD's characteristic symptoms, such as bradykinesia (slowness of movement) and tremors, is a daily challenge. Features that monitor these symptoms and offer feedback may provide crucial information about disease progression and therapy efficacy. Similarly, monitoring sleep quality is critical since many PD patients have sleep difficulties. Wearables that monitor sleep patterns may help people advocate for improved sleep hygiene and enhance their overall health. Finally, including exercise routines in a daily plan is essential to PD care. Wearable fitness trackers may inspire patients, give real-time feedback, and encourage self-care, giving them freedom and control over their health. Creating wearable technologies that meet these demands requires a team effort. Integrating patient opinions throughout the design and development process builds confidence and ensures that technologies meet user demands successfully. This collaborative approach includes data techniques. Patients want explicit transparency about data collection, storage, and utilization. They should be able to quickly get information about these practices and control who has access to their data.

7.8.2 Ethical Considerations

When it comes to treating PD, the incorporation of wearable devices brings up major ethical problems that need to be cautiously considered. Priority one among these issues is the protection of personal information and data. The sensitive health data collected by wearables, such as activity patterns, sleep data, and medication adherence, must be protected to avoid breaches and unauthorized access. In light of the compassionate nature of this information, strong security measures must be implemented to safeguard patient information's confidentiality and prevent any possible damage.

The accuracy of the procedures in identifying symptoms and analyzing data is another critical ethical aspect. In order to guarantee that wearable devices provide dependable and correct data, it is vital to conduct exhaustive validation studies. Erroneous assessments, which might negatively impact patient care and worsen existing health disparities, need to be addressed to prevent them from occurring. Biases in data analysis need to be identified and addressed.

Furthermore, it is necessary to address issues about the approachability and cost-effectiveness of wearable technology. The price of these devices may constitute a barrier to their widespread adoption, especially for those with low incomes and those who live in rural places. If these issues are not addressed, healthcare inequalities could rise, further marginalizing already vulnerable communities.

7.8.3 Multifaceted Solutions

A multifaceted approach to problem-solving: Considering wearable technologies' ethical issues in treating PD requires a diversified approach.

Transparent Data Practices: It is imperative to have effective communication on the collection, storage, and usage of data. Patients should be able to decide who has access to their data and have the right to get information about how their PII is processed. The cultivation of trust between patients and healthcare providers via the use of transparent data procedures helps to guarantee that patients' privacy is protected.

Standardization of Regulations: Clear standards for protecting personal information and data within the healthcare system are essential. Regulations that are standardized ensure that there is uniformity and accountability in the use of wearable devices to track illness. To ensure that patient information is safeguarded, these laws need to include an overview of baseline standards for data encryption, storage, and breach notification systems.

Affordability and Cost-effectiveness: Investigating potential pathways for wearable devices, such as insurance coverage or subsidies, might assist in reducing healthcare inequities and improving accessibility for underserved people. A further benefit of encouraging competition and innovation in the market for wearable devices is that it may result in more inexpensive solutions for patients. To make the advantages of PD management tools accessible to a more significant number of people, regardless of their socioeconomic level, the cost burden that is associated with wearable devices may be reduced.

7.8.4 Conclusion

By continuously monitoring symptoms, tracking medication adherence, and providing essential data points to healthcare providers, wearable devices have the potential to assist individuals who are living with PD develop more effective methods of managing their condition. Additionally, these devices can potentially support the personalized treatment of Parkinson's patients. Nevertheless, it is of the utmost importance to prioritize the requirements of patients and address ethical concerns around the privacy of data, its accuracy, and its accessibility. There is the possibility for wearable technology to revolutionize the care of PD while adhering to ethical principles which might be accomplished by fostering collaboration, establishing explicit legislation, and making the technology accessible to more people. The measures that are being taken to resolve the ethical challenges raise the possibility of a future in which patients will be able to actively manage their disease, which will result in an overall improvement in their quality of life and well-being.

7.9 CASE STUDY 2: THE ETHICAL ISSUES OF PERSONALIZED LANGUAGE MODELS IN BRAIN-COMPUTER INTERFACE COMMUNICATION TECHNOLOGIES: A QUALITATIVE STUDY OF NEUROLOGICAL DISEASE PATIENTS

BCI technology provides an alternate communication approach for those who have lost motor abilities due to degenerative neurological disorders such as ALS, Parkinson's, and multiple sclerosis. BCI communication allows users to pick linguistic units based on brain activity, controlled by neural impulses recorded by external caps or implant chips. Combining NLP with BCI communication systems simplifies utterance generation, reduces cognitive stress, speeds up communication, and reduces the time between thought development and device articulation. However, little information exists on user perception and the value of tailored lexicon incorporation. End-users should be included in the design process, also known as user-centered design, to ensure that technology fits their requirements and is consistent with their beliefs. BCI technology raises ethical concerns, including identity, normality, authority, privacy, and justice. Studying user views on these issues is critical to implementing user-centered design solutions.

To shed light on these ethical concerns, Klein et al. conducted a qualitative study published in the journal *Disability and Rehabilitation: Assistive Technology* in 2022. This study focuses on engaging adults with neurodegenerative disorders to discuss their values toward storing personal lexica inside BCI systems, facilitating meaningful expression as they cope with declining communication abilities (Klein et al., 2022).

7.9.1 Methodology

The technique of purposeful sampling was used in exploratory qualitative research to gather four adult volunteers with various neurological illnesses, including amyotrophic lateral sclerosis (ALS), PD, and multiple sclerosis. Participants were chosen based on predetermined inclusion criteria, which ensured a varied mix of demographics and diseases. Following informed permission, scientists conducted semi-structured interviews with open-ended questions to elicit participants' thoughts and attitudes regarding incorporating individualized language models into BCI communication technology. The interviews lasted an hour and were audio recorded, transcribed verbatim, and deidentified to ensure anonymity.

7.9.2 Results

Individuals with neurodegenerative disorders provided feedback on customizing language models in BCI technology. Overall, respondents recognized the potential advantages of BCI technology and customization but were concerned about privacy, sense of self, and valued connections. Participants' support for customization varied based on their experiences with technology and communication issues. Four primary themes arose from the interviews and questionnaires, including disability, identity, relationships, and privacy. Many participants perceived value in individualized lexicons, emphasizing the significance of keeping identity despite communication obstacles. Nonetheless, they were conscious of the challenges of using personal data while constructing personalized language models (Klein et al., 2022).

7.9.3 Discussion

This study explores the viewpoints of people with neurodegenerative disorders on individualized language models in BCI communication technology. Key results show that prospective users favor enabling devices to keep personal lexica, but they are concerned about privacy, valued connections, and identification. Privacy is an important consideration, with respondents expressing worries about informational, decisional, and bodily privacy. Personalized BCI systems assist in sustaining social closeness and intimacy in valued relationships by enabling users to customize their registers to reflect different social networks. Autonomy is essential because respondents want to fine-tune their social connections in response to losses in other parts of their lives. Users see a customized BCI for communication as beneficial, claiming it improves their capacity to communicate their identity narratively and relationally. User-centered and values-based design should include end-user viewpoints throughout BCI innovation. Addressing these ethical problems is critical to effectively accepting and using BCI communication technology. Despite the recognized limitations, knowing end-user values is still crucial for successful BCI-based communication device development, as it ensures justice and equality in research involving people with complicated communication demands.

7.9.4 Conclusion

Individuals with neurodegenerative disorders must be recognized as key stakeholders in the evolution of NLP in BCI communication technology. The use of individualized language models in these devices raises significant issues of disability, identity, relationships, and privacy. Individuals with neurodegenerative disorders emphasize the usability of BCI communication systems and their congruence with their values and objectives. As a result, qualitative ethics-focused research is an effective method for investigating stakeholder views on new capabilities of BCI communication technologies, notably the management and utilization of individualized language models. This study helped ensure that future advances in BCI communication technology should be ethically sound and sensitive to the needs and concerns of people with neurodegenerative disorders (Klein et al., 2022).

7.10 FUTURE RECOMMENDATIONS FOR ETHICAL AI IMPLEMENTATION

Ethical adoption becomes increasingly crucial as AI breakthroughs revolutionize how doctors or healthcare providers deal with neurodegenerative diseases. Engagement among stakeholders, such as medical practitioners, researchers studying AI, developers, politicians, regulatory bodies, patients, advocacy organizations, and professional organizations, is vital for resolving fairness problems and encouraging justice in AI-powered healthcare. By using multidisciplinary knowledge and incorporating multiple viewpoints, parties may uncover and reduce possible biases in AI algorithms, improving accountability, openness, and confidence in neurodegenerative care. Below are future recommendations for promoting ethical AI implementation in this crucial area (Bhandarkar et al., 2024).

7.10.1 Diversified and Relevant Data Usage

Using varied and relevant datasets throughout AI model design, development, and testing is critical to achieving fair AI-driven neurodegenerative disease management solutions. This technique will allow AI systems to detect, identify, and manage various diseases across several age groups, minimizing gaps and encouraging fairness in healthcare outcomes.

7.10.2 Impartial Audits and Testing of AI Algorithms

Conducting frequent audits and validation procedures for AI algorithms by independent specialists or organizations is vital. These audits assure impartiality and openness, which aids in the detection and elimination of any biases. Building dedicated systems inside medical facilities for algorithmic quality management would enable constant monitoring, bias identification, and appropriate algorithm tweaks to preserve fairness. Education on AI Biases for Clinicians and Patients emphasizes the need for both clinicians and patients to actively engage in ongoing educational initiatives. These efforts establish open talks and feedback mechanisms between healthcare professionals and patients to develop an agreed-upon comprehension of the impact of AI in neurodegenerative disease treatment. Educational processes may be efficiently facilitated via workshops, conferences, and multidisciplinary partnerships.

7.10.3 Enhancing Data Privacy and Security Measures

Improving data privacy and security safeguards is critical to complying with current legislative frameworks such as HIPAA, CCPA, and GDPR. Open methods of communication should be created to inform patients concerning data usage, retention, and dissemination in AI-powered healthcare projects, allowing them to make educated choices about their involvement.

7.10.4 Improving AI Transparency and Explainability

Increasing AI transparency and explainability is of utmost importance. These may be accomplished by developing interpretable algorithms, visualizing decision-making processes, and explaining AI forecasts. Addressing possible biases and limits of XAI would help to avoid overreliance on AI-generated results, fostering trust and responsibility in neurodegenerative disease treatment.

Finally, applying these future suggestions for ethical AI applications in neurological disease management will help build confidence, openness, and inclusion in healthcare systems. By tackling flaws in data and algorithms, parties may guarantee that AI technologies are created and implemented with integrity, effectiveness, and fairness, resulting in better patient outcomes and a more equitable healthcare environment (Markus et al., 2021).

7.10.5 Addressing Ethical Concerns in Clinical Trials and Studies

Hurst et al. (2015) explored the ethical implications of human trials using stem cell-derived dopaminergic neurons to treat PD. The authors emphasized the need to develop explicit ethical principles owing to the complexities of employing human cells produced via sophisticated techniques. They raise ethical concerns, such as informed consent, risk assessment, benefit-risk analysis, and data sharing. The researchers underline the need to properly educate those participating in trials about the experimental component of the therapy, its drawbacks and advantages, and other therapeutic alternatives. They advocate for thorough risk evaluations based on preclinical research and ongoing monitoring during trials. Regarding benefit-risk analysis, they advocate for a detailed examination of therapeutic potential versus potential damage and openness in adverse event reporting and data dissemination to allow learning across trials and organizations. Finally, Hurst et al.'s work emphasizes the need to pay close attention to ethical considerations while developing and executing clinical trials utilizing sophisticated therapeutics such as stem cells or AI. Addressing these issues early on can help build stakeholders' confidence and guarantee that patients get safe and effective therapies while respecting their autonomy and welfare (Hurst et al., 2015).

REFERENCES

Alfalahi, H., Dias, S. B., Khandoker, A. H., Chaudhuri, K. R., & Hadjileontiadis, L. J. (2023). A scoping review of neurodegenerative manifestations in explainable digital phenotyping. *NPJ Parkinson's Disease*, *9*(1), 1–22. https://doi.org/10.1038/s41531-023-00494-0

AlphaFold and beyond. (2023). *Nature Methods*, *20*(2), 163–163. https://doi.org/10.1038/s41592-023-01790-6

Andreotta, A. J., Kirkham, N., & Rizzi, M. (2021). AI, big data, and the future of consent. *AI & Society*, *37*(4), 1715–1728. https://doi.org/10.1007/s00146-021-01262-5

Baines, R., Stevens, S., Austin, D., Anil, K., Bradwell, H., Cooper, L., Maramba, I. D., Chatterjee, A., & Leigh, S. (2024). Patient and public willingness to share personal health data for third-party or secondary uses: Systematic review. *Journal of Medical Internet Research*, *26*, e50421. https://doi.org/10.2196/50421

Bhandarkar, A., Naik, P., Vakkund, K., Junjappanavar, S., Bakare, S., & Pattar, S. (2024). Deep learning based computer aided diagnosis of Alzheimer's disease: A snapshot of last 5 years, gaps, and future directions. *Artificial Intelligence Review*, *57*(2), 1–62. https://doi.org/10.1007/s10462-023-10644-8

Celi, L. A., Cellini, J., Charpignon, M., Dee, E. C., Dernoncourt, F., Eber, R., Mitchell, W. G., Moukheiber, L., Schirmer, J., Situ, J., Paguio, J., Park, J., Wawira, J. G., & Yao, S. (2022). Sources of bias in artificial intelligence that perpetuate healthcare disparities—A global review. *PLoS Digital Health*, *1*(3), e0000022. https://doi.org/10.1371/journal.pdig.0000022

Chatterjee, I. (2021). *Machine learning and its application: A quick guide for beginners*. Bentham Science Publishers.

Chatterjee, I., Agarwal, M., Rana, B., Lakhyani, N., & Kumar, N. (2018). Bi-objective approach for computer-aided diagnosis of schizophrenia patients using fMRI data. *Multimedia Tools and Applications*, *77*(20), 26991–27015. https://doi.org/10.1007/s11042-018-5901-0

Chudzik, A., Śledzianowski, A., & Przybyszewski, A. W. (2024). Machine learning and digital biomarkers can detect early stages of neurodegenerative diseases. *Sensors*, *24*(5), 1572. https://doi.org/10.3390/s24051572

European Parliament and of the Council. (2016). *General Data Protection Regulation*. https://eur-lex.europa.eu/legal-content/EN/TXT/HTML/?uri=CELEX:02016R0679-20160504&qid=1711854411717

Ganesh, S., Chithambaram, T., Krishnan, N. R., Vincent, D. R., Kaliappan, J., & Srinivasan, K. (2023). Exploring Huntington's disease diagnosis via artificial intelligence models: A comprehensive review. *Diagnostics*, *13*(23), 3592. https://doi.org/10.3390/diagnostics13233592

Gichoya, J. W., Thomas, K., Celi, L. A., Safdar, N., Banerjee, I., Banja, J. D., Seyyed-Kalantari, L., Trivedi, H., & Purkayastha, S. (2023). AI pitfalls and what not to do: Mitigating bias in AI. *The British Journal of Radiology*, *96*(1150), 1–8. https://doi.org/10.1259/bjr.20230023

Haider, M., Chauhan, A., Tariq, S., Pathak, D. P., Siddiqui, N., Ali, S., Pottoo, F. H., & Ali, R. (2021). Application of in silico methods in the design of drugs for neurodegenerative diseases. *Current Topics in Medicinal Chemistry*, *21*(11), 995–1011. https://doi.org/10.2174/1568026621666210521164545

Hlávka, J. P. (2020). Security, privacy, and information-sharing aspects of healthcare artificial intelligence. In A. Bohr & K. Memarzadeh (Eds.), *Artificial intelligence in healthcare* (pp. 235–270). Academic Press. https://doi.org/10.1016/b978-0-12-818438-7.00010-1

Hummadi, T., & Chatterjee, I. (2021). An era of brain-computer interface: BCI migration into space. *Neuroscience Research Notes*, *3*(5), 4–12.

Hurst, S. A., Mauron, A., Momjian, S., & Burkhard, P. R. (2015). Ethical criteria for human trials of stem-cell-derived dopaminergic neurons in Parkinson's disease. *AJOB Neuroscience*, *6*(1), 52–60. https://doi.org/10.1080/21507740.2014.990544

Isaak, J., & Hanna, M. J. (2018). User data privacy: Facebook, Cambridge Analytica, and privacy protection. *Computer*, *51*(8), 56–59. https://doi.org/10.1109/mc.2018.3191268

Jumper, J., Evans, R., Pritzel, A., Green, T., Figurnov, M., Ronneberger, O., Tunyasuvunakool, K., Bates, R., Žídek, A., Potapenko, A., Bridgland, A., Meyer, C., Kohl, S. a. A., Ballard, A. J., Cowie, A., Romera-Paredes, B., Nikolov, S., Jain, R., Adler, J.,. . . Hassabis, D. (2021). Highly accurate protein structure prediction with AlphaFold. *Nature*, *596*(7873), 583–589. https://doi.org/10.1038/s41586-021-03819-2

Kenny, L., Moore, K., O' Riordan, C., Fox, S., Barton, J., Tedesco, S., Sica, M., Crowe, C., Alamäki, A., Condell, J., Nordström, A., & Timmons, S. (2022). The views and needs of people with Parkinson disease regarding wearable devices for disease monitoring: Mixed methods exploration. *JMIR Formative Research*, *6*(1), e27418. https://doi.org/10.2196/27418

Khaliq, F., Mahajani, S., Oberhauser, J., & Wakhloo, D. (2023). Decoding degeneration: The implementation of machine learning for clinical detection of neurodegenerative disorders. *Neural Regeneration Research*, *18*(6), 1235. https://doi.org/10.4103/1673-5374.355982

Klein, E., Kinsella, M., Stevens, I., & Fried-Oken, M. (2022). Ethical issues raised by incorporating personalized language models into brain-computer interface communication technologies: A qualitative study of individuals with neurological disease. *Disability and Rehabilitation: Assistive Technology*, *19*, 1041–1051. https://doi.org/10.1080/17483107.2022.2146217

Leginfo.legislature.ca.gov. (2018). *AB-375 Privacy: Personal information: Businesses*. California Legislative Information. https://leginfo.legislature.ca.gov/faces/billTextClient.xhtml?bill_id=201720180AB375

Machado, F. A., & Reppold, C. T. (2015). The effect of deep brain stimulation on motor and cognitive symptoms of Parkinson's disease: A literature review. *Dementia & Neuropsychologia*, *9*(1), 24–31. https://doi.org/10.1590/s1980-57642015dn91000005

Markus, A. F., Kors, J. A., & Rijnbeek, P. R. (2021). The role of explainability in creating trustworthy artificial intelligence for health care: A comprehensive survey of the terminology, design choices, and evaluation strategies. *Journal of Biomedical Informatics*, *113*, 103655. https://doi.org/10.1016/j.jbi.2020.103655

Mersha, M., Lamb, K., Wood, J., AlShami, A., & Kalita, J. (2024). Explainable artificial intelligence: A survey of the need, techniques, applications, and future direction. *Available at SSRN* https://doi.org/10.2139/ssrn.4715286

Munappy, A., Bosch, J., Olsson, H. H., Arpteg, A., & Brinne, B. (2019). *Data management challenges for deep learning*. 2019 45th Euromicro Conference on Software Engineering and Advanced Applications (SEAA), Kallithea, Greece. https://doi.org/10.1109/seaa.2019.00030

Pérez-Toro, P. A., Vásquez-Correa, J. C., Strauss, M., Orozco-Arroyave, J. R., & Nöth, E. (2019). Natural language analysis to detect Parkinson's disease. In K. Ekštein (Ed.), *Text, speech, and dialogue. TSD 2019. Lecture notes in computer science* (Vol. 11697, pp. 82–90). Springer. https://doi.org/10.1007/978-3-030-27947-9_7

Rasheed, K., Qayyum, A., Ghaly, M., Al-Fuqaha, A., Razi, A., & Qadir, J. (2022). Explainable, trustworthy, and ethical machine learning for healthcare: A survey. *Computers in Biology and Medicine*, *149*, 106043. https://doi.org/10.1016/j.compbiomed.2022.106043

Singer, N., & Conger, K. (2019, September 4). *Google is fined $170 million for violating children's privacy on YouTube*. The New York Times. https://www.nytimes.com/2019/09/04/technology/google-youtube-fine-ftc.html

State of California Department of Justice Office of the Attorney General. (2023, May 10). *California Consumer Privacy Act (CCPA)*. State of California—Department of Justice—Office of the Attorney General. https://oag.ca.gov/privacy/ccpa

Stefano, G. B. (2023). Artificial intelligence as a tool for the diagnosis and treatment of neurodegenerative diseases. *Brain Sciences*, *13*(6), 938. https://doi.org/10.3390/brainsci13060938

Williams, N. H. (2023). Artificial intelligence and algorithmic bias. *The international library of bioethics* (Vol. 107). Springer. https://doi.org/10.1007/978-3-031-48262-5_1

Yuste, R., Goering, S., Arcas, B. A., Bi, G., Carmena, J. M., Carter, A., Fins, J. J., Friesen, P., Gallant, J., Huggins, J. E., Illes, J., Kellmeyer, P., Klein, E., Marblestone, A., Mitchell, C., Parens, E., Pham, M., Rubel, A., Sadato, N., … Wolpaw, J. (2017). Four ethical priorities for neurotechnologies and AI. *Nature*, *551*(7679), 159–163. https://doi.org/10.1038/551159a

8 Neurodegenerative Diseases

Alzheimer's Disease

Adejoke Elizabeth Memudu, Baliqis Adejoke Olukade, and Gideon S. Alex

8.1 INTRODUCTION

Alzheimer's disease (AD) is the most predominant dementia (Chen & Yu, 2023), a neurodegenerative disorder with multifactorial risk, characterized by gradual severe loss of cognition and memory (Calvo-Flores Guzmán et al., 2020; Ali et al., 2023; Memudu et al., 2024). AD involves abnormal synaptic activity and brain function (Martín-Belmonte et al., 2022), which precedes phenotypic disease expression and marked neuropathology (John & Reddy, 2021; Ali et al., 2023). AD affects millions of people (Sureda et al., 2020), and it has a profound socioeconomic burden attributed to the lack of effective and efficient treatment (Ezkurdia et al., 2023). The economic burden for managing AD increases direct and indirect economic cost (Wimo et al., 2023). This contributes to making AD a prevalent health condition in the rapidly growing or aging population (Wu et al., 2022), with profound societal implications (Korczyn & Grinberg, 2024) such as loss of work force productivity, and being a leading cause of disability and death globally (Ashrafian et al., 2021; Korczyn & Grinberg, 2024). This age-related health condition is characterized by the formation and accumulation of amyloid-β (Aβ) plaques with intracellular neurofibrillary tangles linked to hyperphosphorylated tau aggregates (Rosselli et al., 2023; Ng et al., 2024), which is wide spread in brain tissue before the phenotypic expression in an individual. As the world's aging population increases, the incidence rate and pathogenic progression of AD also increase within the growing population (Ashrafian et al., 2021; John & Reddy, 2021). Currently, approximately 40 million people live with dementia globally, and it is estimated to increase more than threefold by 2050 (Tiwari et al., 2019). Data reports on dementia in the global population state that "Alzheimer's disease makes up about 70% of all dementia cases and its incidence rate increases with age, doubling every five to ten years for persons between ages 65–85 years and older, with an estimated age-related incidence rate of 0.6%, to 8.4% which implies that AD prevalence increases drastically with advancing age" (GBD 2019 Dementia Forecasting Collaborators, 2022; Baracaldo-Santamaría et al., 2023). These growing statistics of AD negatively impacts the society and economy globally; however, there is growing concern regarding the paucity of information on the socioeconomic impact of AD based on regional and global representation (Jia et al., 2018), and this calls for continued reporting on this aspect of AD in disease data report. AD is a huge societal burden globally (Huang et al., 2023) because of the increasing number of AD cases, lack of effective treatments, and socioeconomic impact (Ahmed & Ishrat, 2020; Huang et al., 2023). It was reported that the estimated total healthcare cost for the treatment of AD in 2020 was estimated at $305 billion, and this cost is predicted to increase to more than $1 trillion as the population ages (Wong, 2020). AD, a common type of dementia, has been reported to be the third-most expensive illness (Liu et al., 2023). This makes dementia to be ranked currently among the

 DOI: 10.1201/9781032711102-8

top 10 global health issues of the 21st century with evident potential to deprive individuals of their dignity, lifestyle, and independence (Villarejo Galende et al., 2021), resulting in a corresponding or ripple effect on affected individual's family or caregiver's daily lifestyle and livelihood (Villarejo Galende et al., 2021). In the last three decades, the global rate of prevalence, incidence, mortality, and disability-adjusted life years (DALYs) has continual increased (Javaid et al., 2021). The monitoring, evaluation, and reporting of social demographic data of dementia's socioeconomic burden give data reports that is directed to provide support and healthcare development strategies (Baracaldo-Santamaría et al., 2023). It is important to discuss AD progression, incidence, and its socioeconomic impact which are significant for policymaking in healthcare resource management and public awareness or outreach engagement for AD and AD-related dementia (Villarejo Galende et al., 2021).

8.2 EPIDEMIOLOGY AND PREVALENCE

Globally, there are about 50 million Alzheimer's patients, and this number is expected to triple by 2050 resulting in risk of people in the workforce being impaired and increasing socioeconomic burden of the disease through the high direct or indirect healthcare cost (Zhang et al., 2021). This projected estimate implies that by the middle of the century, 152 million people worldwide will be living with dementia, out of which the highest incidence will be people in low- and middle-income countries (Zhang et al., 2021). Another epidemiology report hinted that the mortality rate of AD and other dementia is expected to increase from the current 2.4 million to 5.8 million annually by the year 2040 (Javaid et al., 2021). A study reported that Europe and Africa have the largest and lowest incidence, respectively, with women predominantly comprising the number of individuals living with AD (Borumandnia et al., 2020; Zhu et al., 2023). Japan, similar to most Asian countries, has the highest incidence of AD, rising at a pace of 185 per 100,000 people per two years (Borumandnia et al., 2020). American and European countries have the highest trends in yearly increases in AD incidence as compared to the Asian continent (Borumandnia et al., 2020). Another study by Javaid et al. (2021) on Alzheimer's metrics involving 3079 cases per 100,000 has shown that Japan has the highest prevalence, followed by Germany, Italy, Slovenia, Monaco, and Greece; however, when Western Europe was compared to Asia and Africa, the frequency was higher in affluent regions like Western Europe with the significant number of people affected overall throughout South and East Asia, especially in China, Japan, and India. For instance, America had an estimated AD prevalence of 4.5 million in 2000, which is predicted to increase to 13.2 million by 2050 (Harding et al., 2022; Huang et al., 2023).

There has been a steady increase in age-standardized rates (ASR) of incidence, prevalence, mortality, and DALYs for both men and women and women's ASRs were consistently greater than men's (Javaid et al., 2021; Li et al., 2022). Furthermore, a favorable correlation exists between the social demographic index (SDI) and the ASR of incidence, prevalence, and DALYs, where positive correlation is seen between the SDI level and the number of individuals above 70 years old living with AD and other related dementia (Javaid et al., 2021).

8.2.1 Statistics on the Prevalence of AD Globally and Regionally

AD is the fifth leading cause of mortality in people older than 65 years (Wong, 2020), and it is described as the most common debilitating neurodegenerative disorders with increasing incidence and prevalence in the aging population (Carlyle et al., 2021). The prevalence rates of people older than 65 years living with AD was 4%–9%, and women top the chart for nearly every age group (Villarejo Galende et al., 2021). The projected or estimated increase in AD prevalence and incidence is attributed to projected trends in global population growth, aging (Li et al., 2022; Depp et al., 2023; Donnaloja et al., 2023), lower socioeconomic status (SES), and less education (Majoka & Schimming, 2021). A report on AD obtained between 1990 and 2019 revealed an increase incidence and prevalence as 147.95% and 160.84%, respectively (Li et al., 2022; GBD 2019 Dementia Forecasting Collaborators, 2022). A demographic study in the US reported that by the end of 2040, people living with AD might reach 11.2 million, which is about 70% higher than the 2022 report, and this increase will have adverse consequences on the society (Donnaloja et al., 2023). It is noteworthy to mention that various social determinants impact the incidence, prevalence, and risk of AD in the regional and global population (Majoka & Schimming, 2021) among this includes lifestyle, genetic variation, and environmental factors. The projected increase or prevalence of AD varies regionally, with population growth contributing most to the increases in the aging population of sub-Saharan Africa and Asia.

8.2.2 Age-related Risk Factors and Projections for Future Prevalence

Aging is an inevitable risk factor for susceptibility to diseases such as cancer, diabetes, cardiovascular, and neurodegenerative diseases like AD (Korczyn & Grinberg, 2024). The incidence and prevalence of AD as age-related disorder increase rapidly with age (Li et al., 2022; Depp et al., 2023; Korczyn & Grinberg, 2024). About one-third of the older adults population harbor medium to a high margin of AD neuropathology at death, as seen in postmortem brain samples, without significant phenotypic cognitive dysfunction (Carlyle et al., 2021). Aside from aging, other risk factors for the late onset of AD in adults (Baracaldo-Santamaría et al., 2023) include early-life adversity (Majoka & Schimming, 2021), family history of dementia, traumatic brain injury (TBI), genetic factors such as apolipoprotein E (ApoE4 genotype) linked to lipid genes involved in lipid metabolism highly expressed in glial cells (Haney et al., 2024), being female, diabetes, low education level, cardiovascular diseases, obesity, hypertension, environmental factors, and lifestyle factors such as unhealthy diet, poor oral hygiene, and lack of physical exercises (Harding et al., 2022; Khemka et al., 2023). According to Ahmed and Ishrat (2020), the constant increase in the number of AD cases regionally and globally along with the lack of effective treatments is a great problem as it results in poor quality of life for AD patients and an increased mortality rate.

8.3 PATHOPHYSIOLOGY OF ALZHEIMER'S DISEASE

Current persistency in understanding AD pathophysiology following the initial report in 1906 by Alois Alzheimer provides its etiology, which entails a sophisticated interplay of various combinatorial factors such as biological, epigenetic, and environmental toxins such as aluminum and cadmium, and genetic components (Memudu et al., 2020; Nikolac Perkovic et al., 2021; Duara & Barker, 2022). AD has multifactorial etiology, intersecting aging, oxidative stress, and genetic and environmental risk factors (Memudu et al., 2020; Yin, 2023). The neuropathological progression of AD includes cholinergic neuron deficits, neuroinflammation, Aβ-aggregation leading to senile plaques formation, hyperphosphorylation of tau-protein leading to neurofibrillary tangles (Oset-Gasque & Marco-Contelles,

2020; Duara & Barker, 2022), metabolic or bioenergetic deficit, and myelin degeneration (Fernández-Calle et al., 2022; Yin, 2023).

Some of the key pathogenic mechanisms and risk factors for AD are as follows:

8.3.1 Amyloid Plaques and Neurofibrillary Tangles

AD's prominent neuropathological hallmark is the formation of amyloid plaques made up of aggregated Aβ protein and intracellular neurofibrillary tangles of hyperphosphorylated tau protein (Ono & Watanabe-Nakayama, 2021; Baracaldo-Santamaría et al., 2023; Chen & Yu, 2023), evident in postmortem brain samples of AD patients (Busche & Hyman, 2020; Sureda et al., 2020; Ng et al., 2024). Research done using human-induced pluripotent stem cell-derived microglia demonstrates that AD neuropathology relative to fibrillar Aβ formation and tau phosphorylation is attributed to the role of the APOE gene in lipid metabolism (Haney et al., 2024). Aβ is the main neuropathology feature of AD (Congdon et al., 2023). The unstable structural heterogeneity of Aβ molecules is linked to the misfolding and aggregation of Aβ through complex processes triggered by oxidative stress and aging (Ono & Watanabe-Nakayama, 2021; Baracaldo-Santamaría et al., 2023; Chen & Yu, 2023) and results in advancing the progression of AD neuropathology (Ono & Watanabe-Nakayama, 2021; Chen & Yu, 2023). Continuous exposure to mediators of oxidative stress leads to mitochondrial dysfunctions, and neuroinflammation contributes to the progression of tau pathology linked to cognitive deficits (Chen & Yu, 2023) associated with loss of synapses and neurons (Congdon et al., 2023). Current AD drug discovery research is tau-targeted based on the scientific evidence on the interplay between tau pathology and cognitive dysfunction in AD (Chen & Yu, 2023). These therapeutics targeted at tau have promising evidence to slow down AD disease progression and improve cognitive functions (Congdon et al., 2023). The proposed mechanism involves the disaggregation of tau and Aβ molecules into forms such as oligomers, protofibrils, and mature fibrils (Ono & Watanabe-Nakayama, 2021), both within and outside neuron cells (Ashrafian et al., 2021).

8.3.2 Neurotransmitter Dysfunction and Synaptic Loss

The phenotypic hallmark of AD is loss of memory and cognitive function attributed to neurotransmitter dysfunction and loss of synaptic plasticity (Oset-Gasque & Marco-Contelles, 2020). Martín-Belmonte et al. (2022) demonstrated loss of synaptic function in hippocampus due to dysfunction in the metabotropic GABAB receptor signaling through the G protein-gated inwardly rectifying K+ (GIRK or Kir3) channels. GABAergic inhibitory neurons maintain homeostatic synaptic function (Calvo-Flores Guzmán et al., 2020; Martín-Belmonte et al., 2022). This plays a key role in regulating neuronal stimulation mediating neuro cognition such as learning and memory (Martín-Belmonte et al., 2022). However, an increased accumulation of Aβ and tau in the brain synaptic connections alters GABAergic inhibitory functions at synapses (Fernández-Calle et al., 2022; Ali et al., 2023). Acetylcholine has a significant role in cognition and memory potentiation (Allard & Hussain Shuler, 2023). Notable studies on cholinergic neuron dysfunctions mediated by oxidative stress and neuroinflammation results in neurocognitive and memory deficit in AD (Yin et al., 2020; John & Reddy, 2021; Memudu & Adanike, 2022). Cholinergic neurons are very vulnerable to the neurotoxic effects of amyloidosis (Allard & Hussain Shuler, 2023; Berry & Harrison, 2023). Scientific evidence from the forebrain cholinergic neuron has long been implicated in age-related cognitive changes in AD (Berry & Harrison, 2023). Glutamate, an excitatory neurotransmitter of the glutamatergic neuron, works in sync with the GABAergic inhibitory neurons in learning and memory potentiation (Calvo-Flores Guzmán et al., 2020); however, when synaptic junction of neurons are filled with Aβ and tau alters GABAergic inhibitory functions at synapsis (Fernández-Calle et al., 2022; Ali et al., 2023), this results in hyperactivity of glutamatergic neurons resulting in excitotoxicity, neuronal cell death in acute brain injuries such as stroke and trauma. This culminates to exhibiting AD neuropathology and phenotype (Verma et al., 2022). Glutamate mediated neurotoxicity through hyperactivity of the N-methyl-D-aspartate (NMDA) receptors resulting in memory loss and cognitive impairment (Oset-Gasque & Marco-Contelles, 2020). Noradrenaline neurotransmitter has been implicated in neuronal dysfunction in AD. Gutiérrez et al. (2022) reported noradrenergic neurons deficit-linked cognitive dysfunctions in the locus coeruleus of AD patients attributed to low level of noradrenaline in post mortem brain samples of AD patients. The loss of synaptic function and integrity of neurons due to continuous formation and accumulation of Aβ plaques and tau neurofibrillary tangles in the neurons results in cognitive deficits in AD (Pillai & Nadkarni, 2022; Congdon et al., 2023). Synapse dysfunction and loss are early neuropathological characterization of AD (John & Reddy, 2021; Lapresa et al., 2022). Synaptic dysfunction is linked to the accumulation of Aβ (Pillai & Nadkarni, 2022). The interplay between neurotransmitter function and synapses in AD cannot be overemphasized. An *in vivo* noninvasive imaging of neurometabolites shows that abnormal changes in synaptic organization lead to synaptic degradation and neuronal loss, a key factor driving AD pathology (Shahid et al., 2024).

8.3.3 Neuroinflammation

This is a biomarker for the progression of AD neuropathology (Fernández-Calle et al., 2022). It involves the activation of glial cells (microglia and astrocytes) around senile plaques and affected neurons (Wu et al., 2022; Sobue et al., 2023). NMDA-mediated neurotoxicity mediates mitochondrial oxidative damage induced release of microglia-derived factors, a hall mark for neuroinflammation (Oset-Gasque & Marco-Contelles, 2020; Haney et al., 2024). Neurotoxicity induces elevation of the concentrations of Aβ peptides in the brain (Samuels et al., 2023). This initiates astrocytes proliferations (Memudu & Adewumi, 2021; Memudu et al., 2024) and microglial infiltration (Tiwari et al., 2019; Samuels et al., 2023) that mediates phagocytosis, degradation and clearance of Aβ (Qian et al., 2023). Neuroinflammatory response expressed by astrocytes and microglial cells in response to mitochondrial dysfunction and APOE gene-mediated formation of Aβ (John & Reddy, 2021; Fernández-Calle et al., 2022) are used as indicators for the diagnosis and progression of AD. Recently conducted genome-wide association studies (GWAS), identified a large section of AD risk genes enriched in microglia (Sobue et al., 2023). Increasing evidence suggests that both synaptic loss and neuroinflammation constitute early pathologic hallmarks of AD (Wijasa et al., 2020).

8.3.4 Myelin Degeneration

One of the effects of aging on the brain is loss of oligodendrocytes structure and integrity of myelin sheaths, an upstream risk factor for Aβ deposition (Depp et al., 2023;

Sobue et al., 2023). Age-dependent structural anomaly in myelin promotes extracellular β-amyloid deposition as neuritic plaques (Long & Holtzman, 2019; Depp et al., 2023). Therapeutics targeted at improving oligodendrocyte health and myelin integrity promises to be effective to delay the development and slow progression of AD (Depp et al., 2023).

8.3.5 Genetics as Risk Factor

AD is influenced by both genetic and environmental factors, which together hasten its process (Hajjo et al., 2022). Genetics accounts for the largest fraction of the total AD phenotypes (Qian et al., 2023), though it acts heterogeneously and sporadically (Sureda et al., 2020). Genetic deficiencies contribute to AD as mutation in germline, mitochondrial DNA, and single-DNA nucleotide polymorphisms (Oliver & Reddy, 2019). More importantly, an estimation of more than 600 genes contribute to AD, while other contributions are from environment and epigenetics (Cacabelos et al., 2014). Genes like APOE, TREM2 (Nguyen et al., 2020), and SHIP-1 deletion in microglia (Samuels et al., 2023) influence beta-amyloid deposition in AD (Knopman et al., 2021; Zetterberg & Bendlin, 2021; Jansen et al., 2022). Mutation in INPP5D gene expressed in the microglia is a risk factor for developing late-onset AD because it is involved in coding SH2-domain-containing inositol phosphatase SHIP-1 (Samuels et al., 2023). According to Sobue et al. (2023), research conducted via GWAS showcased a large section of AD risk genes richly found in microglia and involved in immune responses in AD pathology. Apo E, the major lipid and cholesterol carrier in the CNS (Fernández-Calle et al., 2022) with three polymorphisms (apoE2, apoE3, and apoE4) impacts AD pathology through synaptic dysregulation, Aβ, tau pathology, and neuroinflammatory response through astrocytes and microglia (Fernández-Calle et al., 2022). The dominant risk factor for late-onset AD is APOE4 gene variant (Blanchard et al., 2022; Fernández-Calle et al., 2022; Yin et al., 2023; Millet et al., 2024). Genetic mutations in amyloid precursor protein (APP), presenilin 1 (PS1), and PS2 cause a small proportion (1%–2%) of early-onset AD (Khemka et al., 2023). The overexpression of APP in the AD brain is caused by the binding of APP intracellular domain (AICD) to Fe65 protein via the C-terminal Fe65-PTB2 interaction, with resultant secretion of Aβ (Iyaswamy et al., 2023). Furthermore, multiple variants of ATP-binding cassette transporter A7 (ABCA7) gene are important risk factors for AD (Qian et al., 2023). ABCA7 variants have the ability to alter ABCA7 protein expression and structure, resulting in abnormal protein functions in lipid metabolism, APP processing, and immune cell function (Qian et al., 2023). Although there are reports that the relationship between lipid metabolism in glia and AD pathology remains poorly understood, genetic risk factors for AD implicate lipid genes involved in lipid metabolism highly expressed in glial cells (Haney et al., 2024). Research shows that ABCA7 deficiency induces neuronal apoptosis through endoplasmic reticulum stress mediated via the PERK/eIF2α pathway resulting in the elevation of Aβ production by upregulating the SREBP2/BACE1 pathway and promoting APP endocytosis (Qian et al., 2023). A study by Haney et al. (2024) using human-induced pluripotent stem cell-derived microglia induced triglyceride synthesis and accumulation of lipids in an APOE-dependent manner, leading to tau phosphorylation and the formation of Aβ. A study showed that SHIP-1 deletion in microglia altered microglial gene expression resulting in infiltration of microglia to Aβ plaques to improve neuronal health by enhanced microglial plaque containment and Aβ engulfment (Samuels et al., 2023).

8.3.6 Environmental and Experiential Factors

Early childhood exposure to stress or high level of stress is associated with increased likelihood of developing AD-related dementia (ADRDs) (Majoka & Schimming, 2021). For instance, early life adversity (ELA) significantly alters brain function later in life and contributes to the pathogenesis of AD (Majoka & Schimming, 2021; Huang et al., 2023). ELA induces neurodegeneration through dysregulation of the hypothalamus-pituitary-adrenal axis, altered gut microbiome, neuroinflammation, oligodendrocyte dysfunction, hypomyelination, and aberrant adult hippocampal neurogenesis which synergistically contributes to cognitive impairment later in life (Huang et al., 2023).

8.3.7 Traumatic Brain Injury

TBI is the largest non-genetic, non-aging associated risk factor for AD (Shin et al., 2021). The number of TBIs from closed head trauma (CHT) significantly increases the risk of developing AD later in life (VanItallie, 2019).

8.3.8 Periodontitis

Periodontitis and AD exist globally in the adult population (Harding et al., 2022). A compromised dentition or dental occlusion could pose a significant risk factor for dementia, as dental occlusion correlates with the presence of APOE4 allele (Popovac et al., 2020). Infectious diseases in the environment due to oral microbiota lead to the risk of developing AD (Sureda et al., 2020). The risk of AD incidence doubles within 10 years from the time of periodontal disease diagnosis; this could be due to advancement in age and a decline in immune protection which negatively impact periodontitis and risk for the onset of AD (Harding et al., 2022; Weintraub et al., 2023).

8.4 CLINICAL MANIFESTATION

AD cognitive dysfunction is related to symptoms such as impairments in language, short-term and long-term memory loss, and behavioral impairments such as aggregation and anxiety (Knopman et al., 2021; Zetterberg & Bendlin, 2021).

8.4.1 Cognitive Decline and Memory Impairment

Patients with AD clinically manifest a progressive dysfunction of cognitive abilities and loss of episodic memory (Congdon et al., 2023). The diagnosis and pathophysiology of AD primarily involve the utilization of cognitive decline and memory impairment (Hajjo et al., 2022; Pillai & Nadkarni, 2022; Congdon et al., 2023). The fundamental use of sophisticated diagnostic measures such as structural imaging aids in the recognition of brain areas associated with AD from irregular patterns in the hippocampal region of the atrophy in the parietal regions (de Souza et al., 2014). The reasons for cognitive decline from changes in the brain are debated, and damage to neurons and synapses involved in memory, learning, and cognitive functions leads to cognitive decline (Huang et al., 2023). An appropriate diagnosis of the disease is confirmed through postmortem examination to show a purposive evidence of neurofibrillary tangles and senile plaques histologically

(Busche & Hyman, 2020; Ng et al., 2024). Multiple cognitive deficits developments have been categorized as hallmark features of AD, and patients with these features also show impaired learning of new information, problem in language, and attentional disturbances (Jansen et al., 2022).

8.4.2 Behavioral and Psychological Symptoms

Cognitive dysfunction in AD individuals calls for the exhibition of some phenotypes of psychological symptoms such as depression, apathy, aggression, agitation, anxiety, and withdrawal from social environment (Knopman et al., 2021; Zetterberg & Bendlin, 2021). AD burden for caregivers and families is featured in psychological changes (Canevelli et al., 2013). The onset of cognitive decline from previous studies showed 80% of individuals with dementia raised at least one phenotype of behavioral and psychological symptoms when compared to mild cognitive impairment (MCI) seen in 50% of patients (Lyketsos et al., 2002). In addition, 75% of these individuals on clinical test prior showed at least one behavioral and psychological symptom of dementia (Steffens et al., 2005) compared to the occurrence of symptoms in MCI patients of 43% of the total population (Lyketsos et al., 2002). The correlation of aggression, apathy, and depression like symptoms in AD individuals with age and cognitive decline showed higher prevalence of aggressive behavior in males compared to females (Lövheim et al., 2009; Zhao et al., 2016).

8.5 IMPORTANCE OF UNDERSTANDING AD'S SOCIETAL IMPACT

The impact of AD on patient's family and society due to financial costs is very significant (Wong, 2020, Dauphinot et al., 2022). Nevertheless, the absence of precision or personalized interventions for the management of AD poses a significant challenge to both the economy and healthcare systems (Skaria, 2022). This results in rising healthcare expenditures in developed countries (Dauphinot et al., 2022). In developed countries, the cost of dementia treatment in individuals with AD in the United State is supposed to increase from $321 billion in 2022 to over $1 trillion by 2050. Two-thirds of these payments are covered by government health taxpayers such as Medicaid and Medicare, while other advanced payments are procured by patients and or their families (Alzheimer's Disease Facts and Figures, 2021). Moreover, with the aging global population, the cases of dementia are projected to increase threefold from 50 million to 152 million by 2050 (Scheltens et al., 2021). The expected global VSL-based economic cost of AD in 2019 was $2.8 trillion, and this cost is predicted to rise to $16.9 trillion in 2050, $8.5 trillion in 2040, and $4.7 trillion in 2030. In 2050, low- and middle-income countries bear 65% of the global value of statistical life based on economic burden. By 2050, 92% of the burden in low and low middle income countries (LMICs) and 60% of the burden worldwide will fall on upper-middle-income countries (Nandi et al., 2022). Life expectancy in AD patients ranges from four to eight years after diagnosis, though some people live up to 20 years . This extended disease duration allows the pathophysiological state to persist for a long time before death, eading to significant economic and societal impacts. Moreover, the most challenging consequences of the disorder are often shared by caregivers (Deb et al., 2017; Alzheimer's Disease Facts and Figures, 2021).

8.6 SOCIOECONOMIC IMPACT

AD imposes a huge burden on society, affecting millions of people globally (Baracaldo-Santamaría et al., 2023; Huang et al., 2023). AD's global socioeconomic position is strongly related to cognitive decline in the society (Frank et al., 2021). The growing global aging population has become a public health concern due to the increasing rate of neuropathological progression and the significant epidemiological and socioeconomic impact, compounded by the lack of effective treatments (Schreiner et al., 2022; Ezkurdia et al., 2023). An estimated 139 million individuals worldwide could be facing various forms of dementia by the year 2050, thereby posing a significant challenge to both healthcare systems and economies (Hampel et al., 2017). AD impacts individuals and their families (Lapresa et al., 2022) because it increases socioeconomic factors such as low income, lack of community-based long-term care services, unemployment, and low education (Gresenz et al., 2020). AD has been associated with great healthcare and non-healthcare resource consumption (Gómez Maldonado et al., 2024). The total cost accrued in AD healthcare includes medical care costs from nursing home care, medications, physician visits, and hospitalizations in addition to non-medical care costs from support caregivers; other costs include intangible costs pertaining to the psychology of the pain and suffering endured by patients and families based on the deterioration of patient and caregiver quality of life (Kopel et al., 2023). There are limited regional reports on socioeconomic factors, family stigma, and their psychological impact on early-onset dementia caregivers (Velilla et al., 2022). The provision of quality healthcare for dementia patients is generally quite expensive (Velandia et al., 2022), and it cannot meet up with the incidence of the AD progression. This corresponds to an increased socioeconomic impact that threatens the sustainability of the healthcare system (Ahmed & Ishrat, 2020). Understanding of the socioeconomic impact of AD is required for appropriate management of healthcare resources and raising social awareness (Villarejo Galende et al., 2021). A majority of patients affected by the disorder lose their source of livelihood and cognitive performance. The degenerative and incapacitating nature of this condition, accompanied by the lack of disease-modifying medications, places a significant financial and societal strain on the healthcare system. The following are affected due to socioeconomy of AD.

8.6.1 Healthcare Costs Associated with AD Diagnosis and Management

AD has huge healthcare costs worldwide, and this strains the healthcare systems and families alike (Wong, 2020). The research costs to develop effective treatments for AD (Iyaswamy et al., 2023; Korczyn & Grinberg, 2024) are very expensive. Earlier, the care of 35 million patients with dementia costs over \$600 billion per year, which is approximately 1% of global gross domestic product (Marešová et al., 2015). In 2020, it is reported to have an estimated total healthcare cost of \$305 billion, with the cost expected to increase to more than \$1 trillion as the population ages (Wong, 2020). The estimation of the total societal costs demonstrates the wide impact of dementia and its main direct and indirect economic components (Wimo et al., 2023). Global economic costs of dementia were estimated to reach US \$1313.4 in 2019, with 61% of people with dementia living in low- and middle-income countries, whereas 74% of the costs occur in high-income countries. The impact of informal care accounts for about 50% of the global costs (Wimo et al., 2023). According to Jia et al. (2018), the socioeconomic costs of AD in China and its impact on global economic burden were calculated from data of 3098 patients with AD in 81 representative centers across China in 2015. It was deduced that the annual socioeconomic cost per patient was US $19,144.36, and the total costs were US $167.74 billion in 2015. The annual total costs are predicted to reach US $507.49 billion in 2030 and US $1.89 trillion in 2050, while the global estimates of costs for dementia were US $957.56 billion in 2015 and will be US $2.54 trillion in 2030 and US $9.12 trillion in 2050, much more than the predictions by the World Alzheimer Report 2015. China bears a heavy economic burden of AD costs, which greatly changes the estimates of AD cost worldwide (Jia et al., 2018). The projected elevation in healthcare cost to compensate for the growing rates of dementia is of critical concern not just in high income countries – where costs associated with care are currently the highest (Wimo et al., 2017). As diagnosis and treatment rates increase along with increases in populations, AD cases will be a challenge to healthcare systems globally (Velandia et al., 2022).

8.6.2 Burden on Caregivers and Family Members

Dementia health burden in the 21st century does not only rob individuals of their dignity and independence, it also has a ripple effect on the affected individual's family (Ahmed & Ishrat, 2020). Patients at advanced stages of AD incur a significant care cost from medical and non-medical care as well as informal caregiving costs; the indirect costs due to reduced productivity of patients and caregivers are also substantial, in addition to intangible costs (Wimo et al., 2016). Caregiving role would impact their individual's personal life and career obligations which causes changes to their daily lives (Kokorelias et al., 2021). The impact on the quality of life of both patients and their caregivers is not readily translated into monetary costs. Concerns have recently been raised that researchers and policymakers are not comprehensively capturing and describing the full scope and magnitude of the socioeconomic burden of AD (Jönsson et al., 2017). Family stigma related to conditions such as having to care for AD parents or loved ones increases the burden among family caregivers. They are the only ones in charge of managing the behavior of their relatives with dementia mostly in daughter caregivers of AD patients who perform more caregiving tasks, are more susceptible to abandon the labor market to care for their parents (Werner et al., 2020). AD caregivers' unemployed status negatively affects caregiver's physical health as a non-working carer, resulting in higher levels of caregiver burden and depression for caregiver (Socci et al., 2021). A rapid decline in financial capacity of caregiver is a hallmark of AD patients with early-stage AD and their caregivers. It places households at significant risk for large adverse changes in liquid assets and reduced net wealth attributed to the cost of AD making most families (Wong, 2020) financially vulnerable during the disease's early stage that impacts the economic status of the afflicted individuals and their families negatively (Gresenz et al., 2020; Wong, 2020). Patients with AD family caregivers are at higher socioeconomic risk to family stigma, higher risk of financial burden, and a reduction in quality of life in terms of energy fatigue and emotional well-being (Villarejo Galende et al., 2021; Velilla et al., 2022). Caregiver distress is ultimately more strongly associated with behavioral disturbances rather than cognitive or functional deterioration.

8.6.3 Functional Decline and Impact on Daily Living

High burdens are presented to patients, families, and society at large in AD conditions as excess in the cost of care and continuous transition in care based on disease stages

(Hajjo et al., 2022). Yet, no drug from the numerous research on AD drug discovery is anti-AD drug specific. However, researchers continue to seek promising anti-AD drugs, though they keep failing at clinical trial levels (Anand et al., 2014; Zheng et al., 2015), which creates more advances in the discussion of therapeutics for AD among researchers. There is currently no established remedy for AD and no recognized therapeutic interventions (Daviglus et al., 2010). In the realm of AD, the decline in functional abilities is thought to stem from gradual impairments in cognitive, emotional, and physical capacities.

8.6.4 Workforce Implications and Productivity Losses

The informal caregiver represents the unpaid family members and friends who give long-term care and support for people living with AD, and this results in some loss in their daily productivity at work place (Frahm-Falkenberg et al., 2016). They have a higher risk of unemployment associated with loss of job because of missing many workdays or forced to reduce working hours to care for their loved ones (Jang & Chen, 2024). Employment rates and income were lower in patients with AD, similar to their partners or caregivers with a lower employment rate and income (El-Hayek et al., 2019). According to Evans (2019), giving the role and contribution of older people in the workforce, it is important to identify and support workers with declining cognitive function at the early onset of AD, and this will create a support system for them in workplace. The rate of unemployment and income is lower in patients with AD, similar to their partners or caregivers (El-Hayek et al., 2019). It is important to mention that AD can cause functional disability and loss of independence (Tahami Monfared et al., 2022) which affects productivity at work, resulting in premature retirement or numerous sick leave (Moschetti et al., 2015; El-Hayek et al., 2019). Caregivers and AD patients experience productivity loss due to missed time from work and reduced performance owing to employees not being fully functional while physically present on the job (Kokorelias et al., 2021; Socci et al., 2021).

8.7 REPORTS ON REGIONAL SOCIOECONOMIC IMPACT OF AD

A report on the estimated socioeconomic burden of AD patients in Spain aged 79.1 ± 7.4 years with a period of diagnosis within 5.8 ± 4.1 years shows that the mean annual cost per patient was estimated at €42,336.4 (Gómez Maldonado et al., 2024). The estimated cost of treating AD in the United States in the year 2022 was estimated at $321 billion; this cost is predicated to surpass $1 trillion by 2050 (Skaria, 2022). The economic and societal impact of chronic disease is increased by indirect expenses like decreased quality of life, lost productivity, and a growing reliance on unpaid informal care from family caregivers (Skaria, 2022). A cross-sectional study shows that 47.3% of AD caregivers reported intense burden, 86.4% significant stress, 57% severe anxiety, and 36.9% mild depression symptoms (Manzini & Vale, 2020).

8.8 PUBLIC HEALTH STRATEGIES TO MITIGATE SOCIOECONOMIC IMPACT OF AD

8.8.1 Awareness Campaigns and Education Initiatives

AD causes marked poor productivity in the workforce, productivity, and individual independence (Tahami Monfared et al., 2022). There is need for massive public healthcare advocacy for people to understand AD clinical presentation through a comprehensive resource training programs directed to educate person on prevention, early diagnosis, treatment, and management for people living with AD in order to minimize the socioeconomic burden of dementia on caregivers (Villarejo Galende et al., 2021).

8.8.2 Supportive Services for Caregivers and Patients

There is need for an adequate support system for the health sector and people living with dementia (Li et al., 2022). Healthcare facilities and governments need to develop funding strategies for early diagnosis and treatment plans for AD (Wong, 2020). The creation of healthcare service homes to cater for people living with AD will reduce the burden of caregivers, which will reduce unemployment and loss of productivity (Norins, 2019).

8.8.3 Policy Recommendations for Improving AD Care and Research Funding

The impact of societal policy engagement that will improve research outcome and early diagnosis of AD (Majoka & Schimming, 2021) is a prerequisite to alleviate burden through health development policies suggested by Li et al. (2022). There is a need for policymakers to effectively budget and prioritize healthcare needs and research funding for AD (Gresenz et al., 2020). There is a need to educate caregivers and vulnerable individuals to healthy lifestyle that reduces the risk of early onset of dementia (Velandia et al., 2022). Continued support for research funding is essential for developing therapeutic strategies for AD therapy (Oset-Gasque & Marco-Contelles, 2020). Extensive research has resulted in no cure for what is perceived as the most common dementing illness: Alzheimer's disease (Korczyn & Grinberg, 2024). Current medications targeting Aβ or tau proteins are unable to reverse AD progression (Liu et al., 2023).

8.9 CONCLUSION

AD is an age-related cognitive and imposes a substantial societal burden attributed to the increasing prevalence of AD in aging populations worldwide due to continuous exposure to risk factors without expected effective therapeutics to mitigate AD neuropathology progression resulting in a strain in healthcare systems, emotional challenges for families, and socioeconomic impacts on communities. However, addressing the socioeconomic burden of AD requires a multifaceted approach involving healthcare policy reforms, increased funding for research and caregiving support, and community-driven initiatives to promote awareness and reduce stigma.

REFERENCES

Ahmed, H. A., & Ishrat, T. (2020). The brain AT2R-A potential target for therapy in Alzheimer's disease and vascular cognitive impairment: A comprehensive review of clinical and experimental therapeutics. *Molecular Neurobiology*, *57*(8), 3458–3484. https://doi.org/10.1007/s12035-020-01964-9

Ali, A. B., Islam, A., & Constanti, A. (2023). The fate of interneurons, GABAA receptor sub-types and perineuronal nets in Alzheimer's disease. *Brain Pathology (Zurich, Switzerland)*, *33*(1), e13129. https://doi.org/10.1111/bpa.13129

Allard, S., & Hussain Shuler, M. G. (2023). Cholinergic reinforcement signaling is impaired by amyloidosis prior to its synaptic loss. *The Journal of Neuroscience: The Official Journal of the Society for Neuroscience*, *43*(42), 6988–7005. https://doi.org/10.1523/JNEUROSCI.0967-23.2023

Alzheimer's disease facts and figures. (2021). *Alzheimers Dement*, *17*(3), 327–406. https://doi.org/10.1002/alz.12328

Anand, R., Gill, K. D., & Mahdi, A. A. (2014). Therapeutics of Alzheimer's disease: Past, present and future. *Neuropharmacology*, *76 Pt A*, 27–50. https://doi.org/10.1016/j.neuropharm.2013.07.004

Ashrafian, H., Zadeh, E. H., & Khan, R. H. (2021). Review on Alzheimer's disease: Inhibition of amyloid beta and tau tangle formation. *International Journal of Biological Macromolecules*, *167*, 382–394. https://doi.org/10.1016/j.ijbiomac.2020.11.192

Baracaldo-Santamaría, D., Avendaño-Lopez, S. S., Ariza-Salamanca, D. F., Rodriguez-Giraldo, M., Calderon-Ospina, C. A., González-Reyes, R. E., & Nava-Mesa, M. O. (2023). Role of calcium modulation in the pathophysiology and treatment of Alzheimer's disease. *International Journal of Molecular Sciences*, *24*(10), 9067. https://doi.org/10.3390/ijms24109067

Berry, A. S., & Harrison, T. M. (2023). New perspectives on the basal forebrain cholinergic system in Alzheimer's disease. *Neuroscience and Biobehavioral Reviews*, *150*, 105192. https://doi.org/10.1016/j.neubiorev.2023.105192

Blanchard, J. W., Akay, L. A., Davila-Velderrain, J., von Maydell, D., Mathys, H., Davidson, S. M., Effenberger, A., Chen, C. Y., Maner-Smith, K., Hajjar, I., Ortlund, E. A., Bula, M., Agbas, E., Ng, A., Jiang, X., Kahn, M., Blanco-Duque, C., Lavoie, N., Liu, L., Reyes, R., … Tsai, L. H. (2022). APOE4 impairs myelination via cholesterol dysregulation in oligodendrocytes. *Nature*, *611*(7937), 769–779. https://doi.org/10.1038/s41586-022-05439-w

Borumandnia, N., Majd, H. A., Khadembashi, N., Olazadeh, K., & Alaii, H. (2020). *Worldwide patterns in Alzheimer's disease and other dementias prevalence from 1990 to 2017: A growth mixture models approach.*

Busche, M. A., & Hyman, B. T. (2020). Synergy between amyloid-β and tau in Alzheimer's disease. *Nature Neuroscience*, *23*(10), 1183–1193. https://doi.org/10.1038/s41593-020-0687-6

Cacabelos, R., Cacabelos, P., Torrellas, C., Tellado, I., & Carril, J. C. (2014). Pharmacogenomics of Alzheimer's disease: Novel therapeutic strategies for drug development. In Q. Yan (Ed.), *Pharmacogenomics in drug discovery and development. Methods in molecular biology* (Vol. 1175, pp. 323–556). Humana Press. https://doi.org/10.1007/978-1-4939-0956-8_13

Calvo-Flores Guzmán, B., Kim, S., Chawdhary, B., Peppercorn, K., Tate, W. P., Waldvogel, H. J., Faull, R. L., Montgomery, J., & Kwakowsky, A. (2020). Amyloid-beta1-42 -induced increase in GABAergic tonic conductance in mouse hippocampal CA1 pyramidal cells. *Molecules (Basel, Switzerland)*, *25*(3), 693. https://doi.org/10.3390/molecules25030693

Canevelli, M., Adali, N., Voisin, T., Soto, M. E., Bruno, G., Cesari, M., & Vellas, B. (2013). Behavioral and psychological subsyndromes in Alzheimer's disease using the Neuropsychiatric Inventory. *International Journal of Geriatric Psychiatry*, *28*(8), 795–803. https://doi.org/10.1002/gps.3904

Carlyle, B. C., Kandigian, S. E., Kreuzer, J., Das, S., Trombetta, B. A., Kuo, Y., Bennett, D. A., Schneider, J. A., Petyuk, V. A., Kitchen, R. R., Morris, R., Nairn, A. C., Hyman, B. T., Haas, W., & Arnold, S. E. (2021). Synaptic proteins associated with cognitive performance and neuropathology in older humans revealed by multiplexed fractionated proteomics. *Neurobiology of Aging*, *105*, 99–114. https://doi.org/10.1016/j.neurobiolaging.2021.04.012

Chen, Y., & Yu, Y. (2023). Tau and neuroinflammation in Alzheimer's disease: Interplay mechanisms and clinical translation. *Journal of Neuroinflammation*, *20*(1), 165. https://doi.org/10.1186/s12974-023-02853-3

Congdon, E. E., Ji, C., Tetlow, A. M., Jiang, Y., & Sigurdsson, E. M. (2023). Tau-targeting therapies for Alzheimer disease: Current status and future directions. *Nature Reviews Neurology*, *19*(12), 715–736. https://doi.org/10.1038/s41582-023-00883-2

Dauphinot, V., Potashman, M., Levitchi-Benea, M., Su, R., Rubino, I., & Krolak-Salmon, P. (2022). Economic and caregiver impact of Alzheimer's disease across the disease spectrum: A cohort study. *Alzheimer's Research & Therapy*, *14*(1), 34. https://doi.org/10.1186/s13195-022-00969-x

Daviglus, M. L., Bell, C. C., Berrettini, W., Bowen, P. E., Connolly, E. S., Jr., Cox, N. J., Dunbar-Jacob, J. M., Granieri, E. C., Hunt, G., & McGarry, K. (2010). National Institutes of Health State-of-the-Science Conference Statement: Preventing Alzheimer disease and cognitive decline. *Annals of Internal Medicine*, *153*(3), 176–181.

de Souza, L. C., Sarazin, M., Teixeira-Júnior, A. L., Caramelli, P., Santos, A. E., & Dubois, B. (2014). Biological markers of Alzheimer's disease. *Arquivos de Neuro-Psiquiatria*, *72*(3), 227–231. https://doi.org/10.1590/0004-282x20130233

Deb, A., Thornton, J. D., Sambamoorthi, U., & Innes, K. (2017). Direct and indirect cost of managing Alzheimer's disease and related dementias in the United States. *Expert Review of Pharmacoeconomics & Outcomes Research*, *17*(2), 189–202. https://doi.org/10.1080/14737167.2017.1313118

Depp, C., Sun, T., Sasmita, A. O., Spieth, L., Berghoff, S. A., Nazarenko, T., Overhoff, K., Steixner-Kumar, A. A., Subramanian, S., Arinrad, S., Ruhwedel, T., Möbius, W., Göbbels, S., Saher, G., Werner, H. B., Damkou, A., Zampar, S., Wirths, O., Thalmann, M., Simons, M., … Nave, K. A. (2023). Myelin dysfunction drives amyloid-β deposition in models of Alzheimer's disease. *Nature*, *618*(7964), 349–357. https://doi.org/10.1038/s41586-023-06120-6

Donnaloja, F., Limonta, E., Mancosu, C., Morandi, F., Boeri, L., Albani, D., & Raimondi, M. T. (2023). Unravelling the mechanotransduction pathways in Alzheimer's disease. *Journal of Biological Engineering*, *17*(1), 22. https://doi.org/10.1186/s13036-023-00336-w

Duara, R., & Barker, W. (2022). Heterogeneity in Alzheimer's disease diagnosis and progression rates: Implications for therapeutic trials. *Neurotherapeutics*, *19*(1), 8–25. https://doi.org/10.1007/s13311-022-01185-z

El-Hayek, Y. H., Wiley, R. E., Khoury, C. P., Daya, R. P., Ballard, C., Evans, A. R., Karran, M., Molinuevo, J. L., Norton, M., & Atri, A. (2019). Tip of the iceberg: Assessing the global socioeconomic costs of Alzheimer's disease and related dementias and strategic implications for stakeholders. *Journal of Alzheimer's Disease: JAD*, *70*(2), 323–341. https://doi.org/10.3233/JAD-190426

Evans, D. (2019). An exploration of the impact of younger-onset dementia on employment. *Dementia*, *18*(1), 262–281. https://doi.org/10.1177/1471301216668661

Ezkurdia, A., Ramírez, M. J., & Solas, M. (2023). Metabolic syndrome as a risk factor for Alzheimer's disease: A focus on insulin resistance. *International Journal of Molecular Sciences*, *24*(5), 4354. https://doi.org/10.3390/ijms24054354

Fernández-Calle, R., Konings, S. C., Frontiñán-Rubio, J., García-Revilla, J., Camprubí-Ferrer, L., Svensson, M., Martinson, I., Boza-Serrano, A., Venero, J. L., Nielsen, H. M., Gouras, G. K., & Deierborg, T. (2022). APOE in the bullseye of neurodegenerative diseases: Impact of the APOE genotype in Alzheimer's disease pathology and brain diseases. *Molecular Neurodegeneration*, *17*(1), 62. https://doi.org/10.1186/s13024-022-00566-4

Frahm-Falkenberg, S., Ibsen, R., Kjellberg, J., & Jennum, P. (2016). Health, social and economic consequences of dementias: A comparative national cohort study. *European Journal of Neurology*, *23*(9), 1400–1407. https://doi.org/10.1111/ene.13043

Frank, M., Hensel, J., Baak, L., Schramm, S., Dragano, N., Weimar, C., Hoffmann, P., Nöthen, M. M., Erbel, R., Jöckel, K. H., Jokisch, M., & Schmidt, B. (2021). Interaction of Alzheimer's disease-associated genetic risk with indicators of socioeconomic position on mild cognitive impairment in the Heinz Nixdorf Recall Study. *Journal of Alzheimer's Disease: JAD*, *82*(4), 1715–1725. https://doi.org/10.3233/JAD-210244

GBD 2019 Dementia Forecasting Collaborators. (2022). Estimation of the global prevalence of dementia in 2019 And forecasted prevalence in 2050: An Analysis for the Global Burden of Disease Study 2019. *The Lancet. Public Health*, *7*(2), e105–e125. https://doi.org/10.1016/S2468-2667(21)00249-8

Gómez Maldonado, L., de Mora-Figueroa, R., López-Angarita, A., Maravilla-Herrera, P., & Merino, M. (2024). Cost of patients with Alzheimer's disease in Spain according to disease severity. *PharmacoEconomics - Open*, *8*(1), 103–114. https://doi.org/10.1007/s41669-023-00451-w

Gresenz, C. R., Mitchell, J. M., Marrone, J., & Federoff, H. J. (2020). Effect of early-stage Alzheimer's disease on household financial outcomes. *Health Economics*, *29*(1), 18–29. https://doi.org/10.1002/hec.3962

Gutiérrez, I. L., Dello Russo, C., Novellino, F., Caso, J. R., García-Bueno, B., Leza, J. C., & Madrigal, J. L. M. (2022). Noradrenaline in Alzheimer's disease: A new potential therapeutic target. *International Journal of Molecular Sciences*, *23*(11), 6143. https://doi.org/10.3390/ijms23116143

Hajjo, R., Sabbah, D. A., Abusara, O. H., & Al Bawab, A. Q. (2022). A review of the recent advances in Alzheimer's disease research and the utilization of network biology approaches for prioritizing diagnostics and therapeutics. *Diagnostics (Basel)*, *12*(12), 2975. https://doi.org/10.3390/diagnostics12122975

Hampel, H., O'Bryant, S. E., Durrleman, S., Younesi, E., Rojkova, K., Escott-Price, V., Corvol, J. C., Broich, K., Dubois, B., & Lista, S. (2017). A precision medicine initiative for Alzheimer's disease: The road ahead to biomarker-guided integrative disease modeling. *Climacteric*, *20*(2), 107–118. https://doi.org/10.1080/13697137.2017.1287866

Haney, M. S., Pálovics, R., Munson, C. N., Long, C., Johansson, P. K., Yip, O., Dong, W., Rawat, E., West, E., Schlachetzki, J. C. M., Tsai, A., Guldner, I. H., Lamichhane, B. S., Smith, A., Schaum, N., Calcuttawala, K., Shin, A., Wang, Y. H., Wang, C., Koutsodendris, N., … Wyss-Coray, T. (2024). APOE4/4 is linked to damaging lipid droplets in Alzheimer's disease microglia. *Nature*, *628*(8006), 154–161. https://doi.org/10.1038/s41586-024-07185-7

Harding, A., Kanagasingam, S., Welbury, R., & Singhrao, S. K. (2022). Periodontitis as a risk factor for Alzheimer's disease: The experimental journey so far, with hope of therapy. *Advances in Experimental Medicine and Biology*, *1373*, 241–260. https://doi.org/10.1007/978-3-030-96881-6_13

Huang, Z., Jordan, J. D., & Zhang, Q. (2023). Early life adversity as a risk factor for cognitive impairment and Alzheimer's disease. *Translational Neurodegeneration*, *12*(1), 25. https://doi.org/10.1186/s40035-023-00355-z

Iyaswamy, A., Thakur, A., Guan, X. J., Krishnamoorthi, S., Fung, T. Y., Lu, K., Gaurav, I., Yang, Z., Su, C. F., Lau, K. F., Zhang, K., Ng, R. C., Lian, Q., Cheung, K. H., Ye, K., Chen, H. J., & Li, M. (2023). Fe65-engineered neuronal exosomes encapsulating corynoxine-B ameliorate cognition and pathology of Alzheimer's disease. *Signal Transduction and Targeted Therapy*, *8*(1), 404. https://doi.org/10.1038/s41392-023-01657-4

Jang, S., & Chen, J. (2024). National estimates of incremental work absenteeism costs associated with adult children of parents with Alzheimer's disease and related dementias. *The American Journal of Geriatric Psychiatry: Official Journal of the American Association for Geriatric Psychiatry*, *32*(8), 972–982. https://doi.org/10.1016/j.jagp.2024.03.011

Jansen, W. J., Janssen, O., Tijms, B. M., Vos, S. J. B., Ossenkoppele, R., Visser, P. J., Aarsland, D., Alcolea, D., Altomare, D., von Arnim, C., Baiardi, S., Baldeiras, I., Barthel, H., Bateman, R. J., Van Berckel, B., Binette, A. P., Blennow, K., Boada, M., Boecker, H., … Zetterberg, H. (2022). Prevalence estimates of amyloid abnormality across the Alzheimer disease clinical spectrum. *JAMA Neurology*, *79*(3), 228–243. https://doi.org/10.1001/jamaneurol.2021.5216

Javaid, S. F., Giebel, C., Khan, M. A., & Hashim, M. J. (2021). Epidemiology of Alzheimer's disease and other dementias: Rising global burden and forecasted trends. *F1000Research*, *10*, 425.

Jia, J., Wei, C., Chen, S., Li, F., Tang, Y., Qin, W., Zhao, L., Jin, H., Xu, H., Wang, F., Zhou, A., Zuo, X., Wu, L., Han, Y., Han, Y., Huang, L., Wang, Q., Li, D., Chu, C., Shi, L., … Gauthier, S. (2018). The cost of Alzheimer's disease in China and re-estimation of costs worldwide. *Alzheimer's & Dementia: The Journal of the Alzheimer's Association*, *14*(4), 483–491. https://doi.org/10.1016/j.jalz.2017.12.006

John, A., & Reddy, P. H. (2021). Synaptic basis of Alzheimer's disease: Focus on synaptic amyloid beta, P-tau and mitochondria. *Ageing Research Reviews*, *65*, 101208. https://doi.org/10.1016/j.arr.2020.101208

Jönsson, L., Lin, P. J., & Khachaturian, A. S. (2017). Special topic section on health economics and public policy of Alzheimer's disease. *Alzheimer's & Dementia: The Journal of the Alzheimer's Association*, *13*(3), 201–204. https://doi.org/10.1016/j.jalz.2017.02.004

Khemka, S., Reddy, A., Garcia, R. I., Jacobs, M., Reddy, R. P., Roghani, A. K., Pattoor, V., Basu, T., Sehar, U., & Reddy, P. H. (2023). Role of diet and exercise in aging, Alzheimer's disease, and other chronic diseases. *Ageing Research Reviews*, *91*, 102091. https://doi.org/10.1016/j.arr.2023.102091

Knopman, D. S., Amieva, H., Petersen, R. C., Chételat, G., Holtzman, D. M., Hyman, B. T., Nixon, R. A., & Jones, D. T. (2021). Alzheimer disease. *Nature Reviews Disease Primers*, *7*(1), 33. https://doi.org/10.1038/s41572-021-00269-y

Kokorelias, K. M., Naglie, G., Gignac, M. A., Rittenberg, N., & Cameron, J. I. (2021). A qualitative exploration of how gender and relationship shape family caregivers' experiences across the Alzheimer's disease trajectory. *Dementia (London, England)*, *20*(8), 2851–2866. https://doi.org/10.1177/14713012211019502

Kopel, J., Sehar, U., Choudhury, M., & Reddy, P. H. (2023). Alzheimer's disease and Alzheimer's disease-related dementias in African Americans: Focus on caregivers. *Healthcare (Basel, Switzerland)*, *11*(6), 868. https://doi.org/10.3390/healthcare11060868

Korczyn, A. D., & Grinberg, L. T. (2024). Is Alzheimer disease a disease? *Nature Reviews Neurology*, *20*(4), 245–251. https://doi.org/10.1038/s41582-024-00940-4

Lapresa, R., Agulla, J., Bolaños, J. P., & Almeida, A. (2022). APC/C-Cdh1-targeted substrates as potential therapies for Alzheimer's disease. *Frontiers in Pharmacology*, *13*, 1086540. https://doi.org/10.3389/fphar.2022.1086540

Li, X., Feng, X., Sun, X., Hou, N., Han, F., & Liu, Y. (2022). Global, regional, and national burden of Alzheimer's disease and other dementias, 1990-2019. *Frontiers in Aging Neuroscience*, *14*, 937486. https://doi.org/10.3389/fnagi.2022.937486

Liu, G., Zhang, L., Fan, Y., & Ji, W. (2023). The pathogenesis in Alzheimer's disease: TREM2 as a potential target. *Journal of Integrative Neuroscience*, *22*(6), 150. https://doi.org/10.31083/j.jin2206150

Long, J. M., & Holtzman, D. M. (2019). Alzheimer disease: An update on pathobiology and treatment strategies. *Cell*, *179*(2), 312–339. https://doi.org/10.1016/j.cell.2019.09.001

Lövheim, H., Sandman, P.-O., Karlsson, S., & Gustafson, Y. (2009). Sex differences in the prevalence of behavioral and psychological symptoms of dementia. *International Psychogeriatrics*, *21*(3), 469–475.

Lyketsos, C. G., Lopez, O., Jones, B., Fitzpatrick, A. L., Breitner, J., & DeKosky, S. (2002). Prevalence of neuropsychiatric symptoms in dementia and mild cognitive impairment: Results from the cardiovascular health study. *JAMA*, *288*(12), 1475–1483.

Majoka, M. A., & Schimming, C. (2021). Effect of social determinants of health on cognition and risk of Alzheimer disease and related dementias. *Clinical Therapeutics*, *43*(6), 922–929. https://doi.org/10.1016/j.clinthera.2021.05.005

Manzini, C. S. S., & Vale, F. A. C. D. (2020). Emotional disorders evidenced by family caregivers of older people with Alzheimer's disease. *Dementia & Neuropsychologia*, *14*, 56–61.

Marešová, P., Mohelská, H., Dolejš, J., & Kuča, K. (2015). Socio-economic aspects of Alzheimer's disease. *Current Alzheimer Research*, *12*(9), 903–911. https://doi.org/10.2174/156720501209151019111448

Martín-Belmonte, A., Aguado, C., Alfaro-Ruiz, R., Moreno-Martínez, A. E., de la Ossa, L., Aso, E., Gómez-Acero, L., Shigemoto, R., Fukazawa, Y., Ciruela, F., & Luján, R. (2022). Nanoscale alterations in GABAB receptors and GIRK channel organization on the hippocampus of APP/PS1 mice. *Alzheimer's Research & Therapy*, *14*(1), 136.

Memudu, A. E., & Adanike, R. P. (2022). Alpha-lipoic acid reverses scopolamine-induced spatial memory loss and pyramidal cell neurodegeneration in the prefrontal cortex of Wistar rats. *IBRO Neuroscience Reports*, *13*, 1–8.

Memudu, A. E., Pantong, S., & Osahon, R. I. (2020). Histomorphological evaluations on the frontal cortex extrapyramidal cell layer following administration of N-acetyl cysteine in the aluminum-induced neurodegeneration rat model. *Metabolic Brain Disease*, *35*, 829–839. https://doi.org/10.1007/s11011-020-00556-91

Memudu, A. E., & Adewumi, A. E. (2021). Alpha-lipoic acid ameliorates scopolamine-induced memory deficit and neurodegeneration in the cerebello-hippocampal cortex. *Metabolic Brain Disease*, *36*(7), 1729–1745. https://doi.org/10.1007/s11011-021-00720-9

Memudu, A. E., Anzaku, F. A., Jibaniya, G. M., & Adanike, R. P. (2024). Neem leaf supplement ameliorates depressive like behaviour in Alzheimer's disease model in adult male Wistar rats. *Asian Journal of Research in Medical and Pharmaceutical Sciences*, *13*(1), 31–46. https://doi.org/10.9734/ajrimps/2024/v13i1245

Millet, A., Ledo, J. H., & Tavazoie, S. F. (2024). An exhausted-like microglial population accumulates in aged and APOE4 genotype Alzheimer's brains. *Immunity*, *57*(1), 153–170.e6. https://doi.org/10.1016/j.immuni.2023.12.001

Moschetti, K., Barragan, N., Basurto-Dávila, R., Cummings, P. L., Sorvillo, F., & Kuo, T. (2015). Mortality and productivity losses from Alzheimer disease among US adults aged 40 to 64 years, 1999 to 2010. *Alzheimer Disease and Associated Disorders*, *29*(2), 165–168. https://doi.org/10.1097/WAD.0000000000000017

Nandi, A., Counts, N., Chen, S., Seligman, B., Tortorice, D., Vigo, D., & Bloom, D. E. (2022). Global and regional projections of the economic burden of Alzheimer's disease and related dementias from 2019 to 2050: A value of statistical life approach. *eClinicalMedicine*, *51*, 101580. https://doi.org/10.1016/j.eclinm.2022.101580

Ng, B., Vowles, J., Bertherat, F., Abey, A., Kilfeather, P., Beccano-Kelly, D., Stefana, M. I., O'Brien, D. P., Bengoa-Vergniory, N., Carling, P. J., Todd, J. A., Caffrey, T. M., Connor-Robson, N., Cowley, S. A., & Wade-Martins, R. (2024). Tau depletion in human neurons mitigates Aβ-driven toxicity. *Molecular Psychiatry*. https://doi.org/10.1038/s41380-024-02463-2

Nguyen, A. T., Wang, K., Hu, G., Wang, X., Miao, Z., Azevedo, J. A., Suh, E., Van Deerlin, V. M., Choi, D., Roeder, K., Li, M., & Lee, E. B. (2020). APOE and TREM2 regulate amyloid-responsive microglia in Alzheimer's disease. *Acta Neuropathologica*, *140*(4), 477–493. https://doi.org/10.1007/s00401-020-02200-3

Nikolac Perkovic, M., Videtic Paska, A., Konjevod, M., Kouter, K., Svob Strac, D., Nedic Erjavec, G., & Pivac, N. (2021). Epigenetics of Alzheimer's disease. *Biomolecules*, *11*(2), 195. https://www.mdpi.com/2218-273X/11/2/195

Norins, L. C. (2019). Predicted economic damage from a quick, simple Alzheimer's disease cure. *Medical Hypotheses*, *133*, 109398. https://doi.org/10.1016/j.mehy.2019.109398

Oliver, D. M. A., & Reddy, P. H. (2019). Small molecules as therapeutic drugs for Alzheimer's disease. *Molecular and Cellular Neuroscience*, *96*, 47–62. https://doi.org/10.1016/j.mcn.2019.03.001

Ono, K., & Watanabe-Nakayama, T. (2021). Aggregation and structure of amyloid β-protein. *Neurochemistry International*, *151*, 105208. https://doi.org/10.1016/j.neuint.2021.10520

Oset-Gasque, M. J., & Marco-Contelles, J. L. (2020). Tacrine-natural-product hybrids for Alzheimer's disease therapy. *Current Medicinal Chemistry*, *27*(26), 4392–4400. https://doi.org/10.2174/0929867325666180403151725

Pillai, A. G., & Nadkarni, S. (2022). Amyloid pathology disrupts gliotransmitter release in astrocytes. *PLoS Computational Biology*, *18*(8), e1010334. https://doi.org/10.1371/journal.pcbi.1010334

Popovac, A., Mladenović, I., Krunić, J., Trifković, B., Todorović, A., Milašin, J., Despotović, N., & Stančić, I. (2020). Apolipoprotein ε4 allele and dental occlusion deficiency as risk factors for Alzheimer's disease. *Journal of Alzheimer's Disease: JAD*, *74*(3), 797–802. https://doi.org/10.3233/JAD-191283

Qian, X. H., Chen, S. Y., Liu, X. L., & Tang, H. D. (2023). ABCA7-associated clinical features and molecular mechanisms in Alzheimer's disease. *Molecular Neurobiology*, *60*(10), 5548–5556. https://doi.org/10.1007/s12035-023-03414-8

Rosselli, M., Uribe, I. V., Ahne, E., & Shihadeh, L. (2023). Culture, ethnicity, and level of education in Alzheimer's disease. *Neurotherapeutics*, *19*(1), 26–54.

Samuels, J. D., Moore, K. A., Ennerfelt, H. E., Johnson, A. M., Walsh, A. E., Price, R. J., & Lukens, J. R. (2023). The Alzheimer's disease risk factor INPP5D restricts neuroprotective microglial responses in amyloid beta-mediated pathology. *Alzheimer's & Dementia: The Journal of the Alzheimer's Association*, *19*(11), 4908–4921. https://doi.org/10.1002/alz.13089

Scheltens, P., De Strooper, B., Kivipelto, M., Holstege, H., Chételat, G., Teunissen, C. E., Cummings, J., & van der Flier, W. M. (2021). Alzheimer's disease. *Lancet*, *397*(10284), 1577–1590. https://doi.org/10.1016/s0140-6736(20)32205-4

Schreiner, T. G., Tamba, B. I., Mihai, C. T., Lőrinczi, A., Baibarac, M., Ciobanu, R. C., & Popescu, B. O. (2022). Nanoporous membranes for the filtration of proteins from biological fluids: Biocompatibility tests on cell cultures and suggested applications for the treatment of Alzheimer's disease. *Journal of Clinical Medicine*, *11*(19), 5846. https://doi.org/10.3390/jcm11195846

Shahid, S. S., Dzemidzic, M., Butch, E. R., Jarvis, E. E., Snyder, S. E., & Wu, Y. C. (2024). Estimating the synaptic density deficit in Alzheimer's disease using multi-contrast CEST imaging. *PLoS ONE*, *19*(3), e0299961. https://doi.org/10.1371/journal.pone.0299961

Shin, M. K., Vázquez-Rosa, E., Koh, Y., Dhar, M., Chaubey, K., Cintrón-Pérez, C. J., Barker, S., Miller, E., Franke, K., Noterman, M. F., Seth, D., Allen, R. S., Motz, C. T., Rao, S. R., Skelton, L. A., Pardue, M. T., Fliesler, S. J., Wang, C., Tracy, T. E., Gan, L., … Pieper, A. A. (2021). Reducing acetylated tau is neuroprotective in brain injury. *Cell*, *184*(10), 2715–2732.e23. https://doi.org/10.1016/j.cell.2021.03.032

Skaria, A. P. (2022). The economic and societal burden of Alzheimer disease: Managed care considerations. *American Journal of Managed Care*, *28*(10 Suppl), S188–s196. https://doi.org/10.37765/ajmc.2022.89236

Sobue, A., Komine, O., & Yamanaka, K. (2023). Neuroinflammation in Alzheimer's disease: Microglial signature and their relevance to disease. *Inflammation and Regeneration*, *43*(1), 26. https://doi.org/10.1186/s41232-023-00277-3

Socci, M., Principi, A., Di Rosa, M., Carney, P., Chiatti, C., & Lattanzio, F. (2021). Impact of working situation on mental and physical health for informal caregivers of older people with Alzheimer's disease in Italy. Results from the UP-TECH longitudinal study. *Aging & Mental Health*, *25*(1), 22–31. https://doi.org/10.1080/13607863.2019.1667295

Steffens, D. C., Maytan, M., Helms, M. J., & Plassman, B. L. (2005). Prevalence and clinical correlates of neuropsychiatric symptoms in dementia. *American Journal of Alzheimer's Disease & Other Dementias*, *20*(6), 367–373.

Sureda, A., Daglia, M., Argüelles Castilla, S., Sanadgol, N., Fazel Nabavi, S., Khan, H., Belwal, T., Jeandet, P., Marchese, A., Pistollato, F., Forbes-Hernandez, T., Battino, M., Berindan-Neagoe, I., D'Onofrio, G., & Nabavi, S. M. (2020). Oral Microbiota and Alzheimer's disease: Do all roads lead to Rome? *Pharmacological Research*, *151*, 104582. https://doi.org/10.1016/j.phrs.2019.104582

Tahami Monfared, A. A., Byrnes, M. J., White, L. A., & Zhang, Q. (2022). The humanistic and economic burden of Alzheimer's disease. *Neurology and Therapy*, *11*(2), 525–551. https://doi.org/10.1007/s40120-022-00335-x

Tiwari, S., Atluri, V., Kaushik, A., Yndart, A., & Nair, M. (2019). Alzheimer's disease: Pathogenesis, diagnostics, and therapeutics. *International Journal of Nanomedicine*, *14*, 5541–5554. https://doi.org/10.2147/IJN.S200490

VanItallie, T. B. (2019). Traumatic brain injury (TBI) in collision sports: Possible mechanisms of transformation into chronic traumatic encephalopathy (CTE). *Metabolism: Clinical and Experimental*, *100*, 153943. https://doi.org/10.1016/j.metabol.2019.07.007

Velandia, P. P., Miller-Petrie, M. K., Chen, C., Chakrabarti, S., Chapin, A., Hay, S., Tsakalos, G., Wimo, A., & Dieleman, J. L. (2022). Global and regional spending on dementia care from 2000-2019 and expected future health spending scenarios from 2020-2050: An economic modelling exercise. *EClinicalMedicine*, *45*, 101337. https://doi.org/10.1016/j.eclinm.2022.101337

Velilla, L., Acosta-Baena, N., Allen, I., Lopera, F., & Kramer, J. (2022). Analysis of family stigma and socioeconomic factors impact among caregivers of patients with early- and late-onset Alzheimer's disease and frontotemporal dementia. *Scientific Reports*, *12*(1), 12663. https://doi.org/10.1038/s41598-022-16400-2

Verma, M., Lizama, B. N., & Chu, C. T. (2022). Excitotoxicity, calcium and mitochondria: A triad in synaptic neurodegeneration. *Translational Neurodegeneration*, *11*(1), 3. https://doi.org/10.1186/s40035-021-00278-7

Villarejo Galende, A., Eimil Ortiz, M., Llamas Velasco, S., Llanero Luque, M., López de Silanes de Miguel, C., & Prieto Jurczynska, C. (2021). Report by the Spanish Foundation of the Brain on the social impact of Alzheimer disease and other types of dementia. Informe de la Fundación del Cerebro. Impacto social de la enfermedad de Alzheimer y otras demencias. *Neurologia*, *36*(1), 39–49. https://doi.org/10.1016/j.nrl.2017.10.005

Weintraub, J. A., Kaeberlein, M., Perissinotto, C., Atchison, K. A., Chen, X., D'Souza, R. N., Feine, J. S., Ghezzi, E. M., Kirkwood, K. L., Ryder, M., Slashcheva, L. D., Touger-Decker, R., Wu, B., & Kapila, Y. (2023). Geroscience: Aging and oral health research. *Advances in Dental Research*, *31*(1), 2–15. https://doi.org/10.1177/08959374231200840

Werner, P., Shpigelman, C. N., & Raviv, T. L. (2020). Family caregivers' and professionals' stigmatic experiences with persons with early-onset dementia: A qualitative study. *Scandinavian Journal of Caring Sciences*, *34*(1), 52–61. https://doi.org/10.1111/scs.12704

Wijasa, T. S., Sylvester, M., Brocke-Ahmadinejad, N., Schwartz, S., Santarelli, F., Gieselmann, V., Klockgether, T., Brosseron, F., & Heneka, M. T. (2020). Quantitative proteomics of synaptosome S-nitrosylation in Alzheimer'S disease. *Journal of Neurochemistry*, *152*(6), 710–726. https://doi.org/10.1111/jnc.14870

Wimo, A., Guerchet, M., Ali, G. C., Wu, Y. T., Prina, A. M., Winblad, B., Jönsson, L., Liu, Z., & Prince, M. (2017). The worldwide costs of dementia 2015 and comparisons with 2010. *Alzheimer's & Dementia: The Journal of the Alzheimer's Association*, *13*(1), 1–7. https://doi.org/10.1016/j.jalz.2016.07.150

Wimo, A., Jönsson, L., Fratiglioni, L., Sandman, P. O., Gustavsson, A., Sköldunger, A., & Johansson, L. (2016). The societal costs of dementia in Sweden 2012—Relevance and methodological challenges in valuing informal care. *Alzheimer's Research & Therapy*, *8*(1), 59. https://doi.org/10.1186/s13195-016-0215-9

Wimo, A., Seeher, K., Cataldi, R., Cyhlarova, E., Dielemann, J. L., Frisell, O., Guerchet, M., Jönsson, L., Malaha, A. K., Nichols, E., Pedroza, P., Prince, M., Knapp, M., & Dua, T. (2023). The worldwide costs of dementia in 2019. *Alzheimer's & Dementia: The Journal of the Alzheimer's Association*, *19*(7), 2865–2873. https://doi.org/10.1002/alz.12901

Wong, W. (2020). Economic burden of Alzheimer disease and managed care considerations. *The American Journal of Managed Care*, *26*(8 Suppl), S177–S183. https://doi.org/10.37765/ajmc.2020.88482

Wu, S., Zhou, J., Zhang, H., & Barger, S. W. (2022). Serine racemase expression differentiates aging from Alzheimer's brain. *Current Alzheimer Research*, *19*(7), 494–502. https://doi.org/10.2174/1567205019666220805105106

Yin, F. (2023). Lipid metabolism and Alzheimer's disease: Clinical evidence, mechanistic link and therapeutic promise. *The FEBS Journal*, *290*(6), 1420–1453. https://doi.org/10.1111/febs.16344

Yin, S., Ran, Q., Yang, J., Zhao, Y., & Li, C. (2020). Nootropic effect of neferine on aluminium chloride-induced Alzheimer's disease in experimental models. *Journal of Biochemical and Molecular Toxicology*, *34*(2), e22429. https://doi.org/10.1002/jbt.22429

Yin, Z., Rosenzweig, N., Kleemann, K. L., Zhang, X., Brandão, W., Margeta, M. A., Schroeder, C., Sivanathan, K. N., Silveira, S., Gauthier, C., Mallah, D., Pitts, K. M., Durao, A., Herron, S., Shorey, H., Cheng, Y., Barry, J. L., Krishnan, R. K., Wakelin, S., Rhee, J., … Butovsky, O. (2023). APOE4 impairs the microglial response in Alzheimer's disease by inducing TGFβ-mediated checkpoints. *Nature Immunology*, *24*(11), 1839–1853. https://doi.org/10.1038/s41590-023-01627-6

Zetterberg, H., & Bendlin, B. B. (2021). Biomarkers for Alzheimer's disease-preparing for a new era of disease-modifying therapies. *Molecular Psychiatry*, *26*(1), 296–308. https://doi.org/10.1038/s41380-020-0721-9

Zhang, X. X., Tian, Y., Wang, Z. T., Ma, Y. H., Tan, L., & Yu, J. T. (2021). The epidemiology of Alzheimer's disease modifiable risk factors and prevention. *The Journal of Prevention of Alzheimer's Disease*, *8*, 313–321.

Zhao, Q.-F., Tan, L., Wang, H.-F., Jiang, T., Tan, M.-S., Tan, L., Xu, W., Li, J.-Q., Wang, J., & Lai, T.-J. (2016). The prevalence of neuropsychiatric symptoms in Alzheimer's disease: Systematic review and meta-analysis. *Journal of Affective Disorders*, *190*, 264–271.

Zheng, H., Fridkin, M., & Youdim, M. (2015). New approaches to treating Alzheimer's disease. *Perspectives in Medicinal Chemistry*, *7*, 1–8. https://doi.org/10.4137/pmc.S13210

Zhu, Z., Zheng, Z., Zhou, C., Cao, L., & Zhao, G. (2023). Trends in prevalence and disability-adjusted life-years of Alzheimer's disease and other dementias in China from 1990 to 2019. *Neuroepidemiology*, *57*(4), 206–217.

9 Neurodegenerative Disorders
Frontotemporal Dementia, ALS and Huntington's Disease

Dhvanit Rajdeep

9.1 INTRODUCTION

Neurodegenerative disorders are mainly characterized by loss of neurons. Here, we mainly discuss three neurodegenerative disorders that are frontotemporal dementia (FTD), amyotrophic lateral sclerosis (ALS) and Huntington's disease (HD) and how they are connected with each other, and we are also learning about the relationship between these disorders because of the new evidence emerging in recent times which describes the relationship of FTD and ALS with each other and also explaining the relationship of FTD and HD with each other. Recent findings indicate that TDP-43-positive cytoplasmic inclusions are present in almost all cases of ALS, and over 50% of FTD patients have renewed interest in the relationship between these progressive neurodegenerative disorders (Mackenzie & Rademakers, 2008). ALS was previously suspected to be caused by uncommon symptoms such as overt cognitive symptoms and frank dementia, although these symptoms were reported in early descriptions. Nevertheless, certain FTD patients experience the development of ALS. Both disorders are commonly clustered in families, with FTD and ALS being common symptoms or coincidentally occurring. While cognitive and neuropsychological assessments may reveal symptoms that are initially inconspicuous and are often overlooked by doctors, 20%–50% of ALS patients meet the consensus criteria for probable or definite FTD with appropriate measures. Voxel-based morphometry magnetic resonance imaging (MRI) in ALS and FTP-ALS patients has also revealed structural abnormalities that support the overlap between these diseases (Bang et al., 2015; Kiernan et al., 2011).

The defining characteristic of HD is the prevalence of subcortical movements, which are mainly involuntary. Social behavior of patients is influenced, but not as much as in FTD, and there is often marked disintegration in interpersonal relationships. It is common to describe patients as being egocentric, indifferent to sympathy and empathy, and mentally rigid, sometimes with fixed opinions that may conflict with the public view or available evidence. This can be problematic for many of these individuals who struggle to maintain their well-being within an institution. Similarly, FTD deficits have been observed in the emotional processing. Cognitive changes are mainly focused on frontal executive function, but with less significance than FTD, and memory impairment is linked to inefficient encoding and retrieval methods rather than a primary lack of retention (Snowden et al., 2003). Because of the recently founded overlapping of these disorders with each other, we need to understand the basics of the disorders (FTD, ALS and HD), and that's why we discuss the clinical features, imaging methods, pathophysiological mechanisms and ways of management for FTD, ALS and HD (Bang et al., 2015; Kiernan et al., 2011; Walker, 2007).

DOI: 10.1201/9781032711102-9

9.2 FRONTOTEMPORAL DEMENTIA

Currently, FTD includes clinical disorders that contain variations in behavior, language, executive control and motor symptoms (Bang et al., 2015). Here, we will characterize the core FTD spectrum disorders through its clinicopathological variations such as behavioral variant FTD (bvFTD), nonfluent/agrammatic variant primary progressive aphasia (nfvPPA) and semantic variant PPA (svPPA). Disorders related to FTD will be reviewed to understand the relationship with other pathologies including FTD with motor neuron disease (FTD-MND), progressive supranuclear palsy syndrome (PSP-S) and corticobasal syndrome (CBS) (Bang et al., 2015). The term frontotemporal lobar degeneration (FTLD) is utilized for medical conditions that produce the degeneration of frontal and temporal lobes. Different types of therapeutic strategies will also be reviewed for symptom management.

9.2.1 Clinical Features

Three clinical variants of FTD can be identified: behavioral-variant FTD, non-fluent variant primary progressive aphasia and semantic-variant primary progressive aphasia. Imbalanced speech, grammar and word output are also present in propensity for delayed attention due to learning delays caused by neural changes (Rascovsky et al., 2011). The diagnostic standards define traits that contribute to the likelihood of detecting FTD-associated neuropathy (Burrell et al., 2011). Those with the disease in its final stages experience difficulty in eating, moving and swallowing. The onset of symptoms usually leads to death, which is typically caused by pneumonia or other secondary infections and occurs approximately 8 years later (Bang et al., 2015).

9.2.1.1 Behavioral-variant Frontotemporal Dementia

Behavioral-variant FTD is marked by personality changes, disinhibition and apathy, which are early symptoms. Disinhibition of behavior can lead to tactless and socially inappropriate behavior, impulsive and careless behavior and can also result in criminal behavior. People with behavioral variant might also do some embarrassing activities, because of which they get embarrassing personal remarks from their peers. (Bang et al., 2015). Reduction in fiscal decision-making can result in financial ruin due to reduced inhibition (Bang et al., 2015). While patients may make inappropriate sexual comments, they tend to have a lower libido. The symptoms of apathy may include sluggishness in activities like work, hobbies and relationships while also being less interested in hygiene (Neary et al., 1998). A patient who attended the funeral of his spouse was more apprehensive than to go to the grocery store. Consequently, the patient may be provoked to anger family and friends. The stereotypical behaviors of patients are characterized by simple repetitive movements, compulsive ritualistic behavior and the repetitive use of words or phrases (Bang et al., 2015). In behavioral-variant FTD, hyperorality is marked by binge eating, increased consumption of sweets or alcohol and weight gain.

9.2.1.2 Semantic-variant Primary Progressive Aphasia

A condition called semantic dementia was identified by exhibiting semantic aphasia and associative affability. Early asymmetrical degeneration of the anterior temporal lobes and amygdala leads to symptoms. Anomia arises due to semantic loss, which causes problems with word placement and comprehension of words. The left temporal lobe variant is approximately three times more prevalent than the right temporal lobe variant. The patient's ability to comprehend specific words is compromised (Bang et al., 2015). The pronunciation of nouns is typically higher than that of verbs or pronouns. The presence of atypical spelling or pronunciation errors in words such as bouquet and bo-ket is evident in patients with surface dyslexia and dysgraphia. Other language domains are not affected, particularly in the initial stage of the illness, and patients maintain their fluency in speech and grammar (Boxer et al., 2013). In this variant, semantics are lost (in the case of the left temporal lobe), but some functions on the right side (i.e., visual attention) are increased (Seeley et al., 2009). Due to the presence of left temporal lobe variant, patients frequently experience visual stimuli like playing puzzles, jewellery beading, painting and brightly colored objects. This is particularly evident in these patients. Patients with right temporal lobe variants exhibit verbal stimuli through words and symbols.

9.2.1.3 Primary Progressive Aphasia

In patients with primary progressive aphasia, there is an ongoing and inward decline of spoken language at the beginning of the disease. Language impairment is the primary symptom in the initial 2 years of the illness (Gorno-Tempini et al., 2011). Some of the deficits seen are language production, object naming, syntax or word comprehension and are apparent during conversation or through speech and language assessment (Bang et al., 2015). FTD is the most common cause of this condition, but primary progressive aphasia can be linked to Alzheimer's disease. During the first stage of the disease, there should be no behavioral disturbances, as these changes are indicative of onset-dependent FTD (Mesulam, 2001).

9.2.1.4 Non-fluent Variant Primary Progressive Aphasia

Non-fluent primary progressive aphasia is marked by delayed, labor-intensive and slow speech production, as well as grammar errors or misuse (agrammatism) (Bang et al., 2015). This condition is rare. Patients frequently experience speech sound errors, which can include insertions (increases), deletions, substitutions or transpositions and distortions (Olney et al., 2017). It has been found that some patients might have trouble understanding sentences with complex syntactic constructions but retain the ability to understand simpler sentences with the same semantic content (Bang et al., 2015). Mild grammatical errors are discovered in written language production and syntactic comprehension tests during the early stages of the disease (Zhang et al., 2009). Certain patients exhibit no signs of language deficiency, but their writing remains unaffected.

9.2.2 Motor Symptoms

Motor neuron disease is characterized by symptoms such as hyperreflexia, extensor plantar response, spasticity, weakness, muscle atrophy, fasciculations, dysarthria, dysphagia and pseudobulbar, affecting approximately 12.5% of patients with behavioral-variant FTD (Bang et al., 2015). The majority of patients with FTD exhibit mild symptoms, known as motor neurons. Patients with FTD experience early parkinsonism, which is most frequently observed in those with behavioral-variant frontal dementia and those suffering from non-fluent variant primary progressive aphasia. Patients suffering from FTD may also exhibit symptoms of CBS or progressive supranuclear palsy syndrome. This condition is not monogenic. There is a prevalence of behavior change, such as executive dysfunction, impulsivity and omission (Bang et al., 2015).

9.2.3 Imaging for Frontotemporal Dementia

Advances in imaging techniques have made it possible to diagnose FTD using a range of available methods. Atrophy patterns are observed in the frontotemporal region, and if atrophy is present during frontal dementia, it can be further described by structural MRI and computed tomography (CT) (Rosen et al., 2002). The regions display disproportionate hypoperfusion and undereating, as well as abnormalities in fluorodeoxyglucose PET, functional MRI and single-photon-emission CT. This makes it an increasingly active field of research. Due to the lack of amyloid deposition as a neuropathological feature of FTD, amyloid tracer imaging can be used to differentiate between them (Bang et al., 2015).

9.2.4 Genetics of Frontotemporal Dementia

FTD often has a strong genetic constituent adding to its pathogenesis. It has been found that more than half of FTD cases are considered to be sporadic, but up to 40% of cases have a family history of dementia, psychiatric disease or motor symptoms, with at least 10% of cases having an autosomal dominant pattern (Rohrer et al., 2009). The main three genes commonly linked with FTD are *C9ORF72*, *MAPT* and *GRN 9* (Le Ber, 2013).

9.2.5 Management of Frontotemporal Dementia

The treatment for FTD is not based on approved disease-modifying drugs. The primary concern is the management of behavioral symptoms through treatment. The use of selective serotonin reuptake inhibitors can enhance severe compulsiveness, agitation, aggressiveness and impulsivity, as well as aberrant eating behavior (Thompson et al., 2003). Atypical antipsychotics can be administered in small amounts to treat behavior abnormalities (Asmal et al., 2013). The use of atypical antipsychotics in treating elderly patients with dementia should be done with caution. Behavioral abnormalities in individuals with FTD may be intensified by cholinesterase inhibitors, which are not beneficial (Bang et al., 2015). FTD symptoms do not progress or slow down due to the use of memantine.

Advancements in the knowledge of FTD have uncovered new therapeutic targets that could be utilized to create effective drugs to prevent the disease (Chow et al., 1999). By discovering that pathological tau protein spreads in a prion-like manner across anatomical pathways of intrinsic connectivity, researchers have discovered how antibodies could prevent the trans-neuronal spreading and potentially prevent disease progression (Bang et al., 2015). A method that aims to decrease the overall number of pathological tau species is being researched. Neurodegeneration in PGRN mutation carriers is associated with decreased progranular haploin concentration and progenitor cell numbers, so studies are being conducted to control this process using various drugs (Capell et al., 2011; Cenik et al., 2011).

9.3 AMYOTROPHIC LATERAL SCLEROSIS

An idiopathic neurodegenerative disease called Amyotrophic lateral sclerosis can be fatal (Kiernan et al., 2011). Here, we will go through a summary of current knowledge on the cause of ALS, what factors contribute to its development and why all patients of ALS are different from each other in many prospects. These questions have been subjected to painfully slow progress since Charcot first described ALS 150 years ago. We are describing information about ALS such as why research interest in ALS has been stimulated nowadays and clinical heterogeneity problems which have been identified in research related to ALS. We will also be learning about the way that survival is now assumed to differ in several factors such as clinical presentation (phenotype), rate of disease progression, early presence of respiratory failure or the nutritional status of patients with regard to cases of ALS. Information about clinical trials which are trying to develop measures to prevent neurodegeneration for ALS has also been described here (Kiernan et al., 2011).

9.3.1 Pathophysiological Mechanisms of ALS

The development of ALS appears to involve multiple factors, with mounting evidence of an intricate interplay between genetic and molecular pathways. A recent study also highlights the complexity of pathophysiology. A disorder that occurs in the motor neuron system of humans during development, such as ALS, may manifest itself as an adult condition. A toxic gain of function of the SOD1 enzyme is believed to be caused by genetic mutations that generate free radicals, leading to cell injury and death. This is now widely understood. Also, SOD1 mutations result in the creation of intracellular aggregates due to conformational instability and misfolding of a SOD1 peptide, which impairs normal proteosomic function and disrupts important cell functions. The excitotoxicity of glutamate can lead to neurodegeneration by stimulating calcium-dependent enzymatic pathways, particularly when it stimulates the excitatory activity of these postsynaptic receptors. Free radicals, which can damage intracellular organelles and upregulate proinflammatory mediators during excitotoxicity, can also cause neurodegeneration (Kiernan et al., 2011). However, it is not known how glutamate-induced excitotoxicity mediates motorneuronal degeneration in humans. Some studies have found in their results that through transcranial magnetic stimulation, cortical hyperexcitability is an early feature in patients with sporadic ALS and precedes the clinical onset of familial ALS which had provided support for the "dying forward" hypothesis. Some clinical observations were found in these studies such as motor neurons without a monosynaptic connection with corticomotorneurons. Also found were the oculomotor, abducens and Onuf's nuclei, which are typically spared in ALS, and studies also found that pure LMN forms of ALS are rare, whereas subclinical UMN involvement is invariably detected with transcranial magnetic stimulation studies (Kiernan et al., 2011).

9.3.2 Clinical Presentations of ALS

Major manifestations of ALS seen are limb-onset ALS with a combination of UMN and LMN signs in the limbs; bulbar-onset ALS, with speech and swallowing problems; primary lateral sclerosis with pure UMN involvement; progressive muscular atrophy and pure LMN involvement (Gordon et al., 2006). The presence of UMN and LMN features, which involve the brainstem and multiple spinal cord regions of innervation, is clinically significant in ALS. It has been found that about 25% of patients might present with bulbar-onset disease, about 70% with limb-onset disease and about 5% with initial trunk or respiratory involvement, subsequently spreading to involve other regions (Turner et al., 2009). Symptoms may manifest as weight loss, poor prognosis, cramping and fasciculations without muscle weakness, emotional lability or frontal lobe-type cognitive dysfunction. Flaccid dysarthria and later dysphagia are the hallmarks of bulbar LMN dysfunction, which includes tongue wasting, weakness, and fasciculations (Kiernan et al., 2011). These disorders are rare. Nasal speech is a result of flamboyant dysarthria, which causes hoarseness, palatal weakness, and weakened coughing. Some very common symptoms of ALS are fatigue and decreased ability to exercise.

It has been found in pure bulbar palsy phenotype, which is mostly found in women over 65 years old and has UMN features, typically results in an estimated 2- to 4-year survival. In addition, primary lateral sclerosis progress is slower than in the classic form of ALS. The diagnosis of primary lateral sclerosis should be delayed for at least 4 years, as the LMN symptoms may still develop even if the initial presentation is pure spastic syndrome (Kiernan et al., 2011). ALS patients commonly experience dysphagia, which results in weight loss and poor prognosis due to malnutrition. The majority of ALS patients experience respiratory compromise, which causes symptoms such as exertional dyspnea, orthopnea, hypoventilation with resultant hypercapnia and early morning headaches (Leigh et al., 2003). Patients who develop dyspnea at rest are in a state of imminent death. Pneumonia is a common cause of respiratory failure caused by the gradual weakening of the respiratory muscles.

9.3.3 Imaging in ALS

The diagnostic pathway for ALS has been significantly improved by neuroimaging, which has enabled MRI to exclude other pathological causes. Cerebral atrophy was detected using MRI with voxel-based cognitive impairment detection, while patients with familial disease and those who had longer life expectancy differed in their regional focus. MRI can be used to visualize the brain in 3-dimensional configuration, which can also indicate focal abnormality (Kiernan et al., 2011). Magnetic results obtained through magnetic resonance spectroscopy reveal that proton-containing metabolic substances such as N-acetylaspartate have been utilized as neuronal loss markers (Talbot, 2009). ALS patients have a lower primary motor cortex N-acetylaspartate to creatine ratio than controls in some studies, and magnetic resonance spectroscopy's ability to detect UMN dysfunction is particularly effective in distinguishing patients with progressive muscular atrophy from those with ALS (Kiernan et al., 2011).

9.3.4 Management of ALS

Neuroprotective therapy for ALS involves using riluzole, which inhibits the release of glutamate. Patients who received riluzole in two large randomized controlled trials had an improved survival time of 3–6 months. Treatment in specialized multidisciplinary ALS clinics appeared to be more effective than in other settings, with patients with moderate functional impairment receiving the most beneficial benefits. ALS patients who receive non-invasive ventilation experience improved quality of life and survival (Kiernan et al., 2011). Non-invasive ventilation guidelines are based on the combination of symptoms that indicate respiratory muscle weakness (dyspnea and orthopnea), as well as signs of muscular weakness such as overnight saturation on oximetry, increased partial pressure of carbon dioxide (PCO_2) of less than 65 mmHg, reduced forced vital capacity or sniffing nasal inspiratory pressure (Vucic and Kiernan, 2007). Surgical ventilation through tracheostomy is an alternative to non-invasive ventilation in patients with ALS who cannot tolerate it due to progressive respiratory muscle weakness (Kiernan et al., 2011). Despite the fact that invasive ventilation can extend survival time, it is rarely implemented in most countries due to the practical difficulties, cost and significant quality of life loss (Eisen et al., 1992). The combination of motor neurons and central respiratory centers that contribute to the phrenic nerve is indicative of respiratory failure. ALS patients often experience respiratory compromise upon diagnosis. However, they may be better for patients with severe facial muscle weakness.

The prognosis is heavily influenced by malnutrition. A percutaneous gastrostomy tube is recommended for patients who have undergone substantial weight loss, even in the absence of dysphagia, to ensure adequate caloric and fluid intake (Kiernan et al., 2011). The decision to proceed with a gastrostomy should be communicated early on in disease progression as morbidity is more prevalent when vital capacity falls below. Resting energy expenditure reflects an increase in metabolic rate, which is associated with reduced survival. The role of muscle mitochondria in the pathogenesis of ALS is still unknown, and it remains unclear what mechanisms underlie a hypermetabolic state (Dupuis et al., 2009).

9.4 HUNTINGTON'S DISEASE

The phenotype of HD is marked by chorea and dystonia, incoordination, cognitive decline and behavioral difficulties, which are caused by autosomal dominant neurodegenerative disorders (Walker, 2007). Symptoms usually appear in middle age after the person has had children, but symptoms can appear any time from childhood. The N-terminus polyglutamine strand is formed as a result of an expanded CAG repeat, which leads to the development of HD mutant protein called huntingtin (Walker, 2007).

9.4.1 Clinical Features of Huntington's Disease

A combination of neuropsychiatric symptoms, a gradual movement disorder and also dementia characterizes HD (Stoker et al., 2022). The condition can emerge at any point in life, but it progresses through customary phases. With the progression of the disease, this movement disorder can progress from hyperkinetic to hypokinetic and subsequently worsen cognition and mood swings. The Unified Huntington's Disease Rating Scale is a standard tool that assesses these traits in the clinic, using motor, cognitive, behavioral and functional assessments and an independence scale (Stoker et al., 2022). I'm not sure if these are early signs or simply an overestimation of normalcy and the passing of time. Some patients still lack significant symptoms despite apparent changes in their symptoms, which can be attributed to inadequate knowledge of their condition (Aylward et al., 2004). Some patients with heterozygous CAG expansions do not exhibit a consistent pattern of symptoms, which makes the treatment difficult (Stoker et al., 2022).

9.4.2 Movement Disorder

HD is best characterized by the progressive choreiform movements that often occur in early stages of the disease, particularly during one's fourth or fifth decade. It has been found that the age at which patients present with manifest HD closely relates to the CAG repeat length and in late-onset cases having smaller repeat lengths (Andrew et al., 1993). The age of onset is not accurately determined by repeat length at an individual level, but it is likely that more than 55 repeats will result in juvenile HD. The initial phase of hyperkinetic behavior may involve minor adjustments, such as flicking fingers while walking or getting up and lying in bed, but it can become more pronounced and all body parts affected in a symmetrical manner. Motor impersistence is a significant characteristic of patients who are unable to sustain voluntary muscle contractions. Another instance is the "milkmaid's grip", which occurs when the patient is unable to hold onto a firm handshake (Stoker et al., 2022). Despite the fact that chorea development is frequently used to identify the onset of symptoms, the severity of an illness is not directly related to its level. Mild parkinsonism, which occurs before the age of 21, is commonly associated with juvenile HD, including severe bradding and dystonia, as well as occasional tremor and rigidity, but occasionally with later episodes in patients (Stoker et al., 2022). This condition is also common among those aged between 18 and 29 years old.

9.4.3 Imaging for Huntington's Disease

Achieving moderate-to-severe Huntington's diagnosis can be done through routine MRI and CT scans which may show striatal volume loss and increased size of the frontal horns of the lateral ventricles, but scans are generally insufficient for diagnosing early symptoms (Walker, 2007). In fact, with these techniques, caudate atrophy becomes noticeable as early as 11 years ago before the expected onset and putaminal atrophy were identified as early as 9 years (Walker, 2007). People with the HD gene but no indication of its development can develop it as found by clinical or neuropsychological tests conducted over a 2-year period (Walker, 2007). It had been found that tensor-based magnetic resonance morphometry has indicated progressive loss of striatal volume (Walker, 2007).

9.4.4 Management

The investigation of antisense oligonucleotide therapy and other potential treatments has been conducted in HD, but none have demonstrated any decrease in disease progression. Recently, the European Huntington's Disease Network commissioned a publication that published dozens of recommendations for managing HD symptoms (Bachoud-Lévi et al., 2019). We must maintain a positive perspective on treatment and emphasize that there are multiple ways to assist patients, even though there is no cure. Clinical clinician experience is the most important factor in selecting a drug, as there is little data on these medications. To make an informed decision on a drug, it is important to consider the overall symptoms of the patient as some treatments may have unintended side effects (e.g. tetrabenazine can improve chorea and yet provide worse effects in neuropsychiatric conditions). The non-pharmacological aspects of care, including access to appropriate health plans and support services like meals or housing/benefits for patients, must also be considered (Stoker et al., 2022). Physiotherapy and occupational therapy may also be useful, with some suggesting that increased activity levels can help maintain function (Thompson et al., 2013). These patients and their families can receive vital assistance from patient support groups like those mentioned above, which should be sought by patients themselves. The existence of HD and advice on managing can be found through these organizations that provide a wealth of information.

9.4.5 Management of Movement, Neuropsychological and Neuropsychiatric Symptoms

The only drug approved for use in HD chorea treatment is tetrabenazine (Stoker et al., 2022). The uptake of monoamines (including dopamine) into synaptic vesicles is reduced by blocking the vesicular monoamine transporter 2. It is recommended to start with a low dose of 125 mg per day, and then the dose is increased every 60 days. Patients are required to take two to three doses daily, as their doping regimens vary due to the variable half-life. Among the potential benefits it can offer are significant functional impairments caused by chorea (such as serotonin withdrawal) and depression, with minimal neuropsychiatric indications. Suicide is a potential risk, but it should be monitored closely, and the drug should not be used if it occurs. Dosage reduction may result in better healing for other adverse reactions. Those with co-morbid behavioral problems, depression or psychosis are typically given antipsychotic drugs (Sanders et al., 2014). Some of the side effects found during trials include sedation, dry mouth, metabolic syndrome and weight gain (indeed, it may be preferred in part because it can help to reduce weight loss and improve sleep and mood) (Stoker et al., 2022). Olanzapine has been ineffective or unreliable, so there is no need to treat it with another neuroleptic (Burgunder et al., 2011).

Neuropsychological tests are routinely conducted to identify the nature of the patient's deficits, enabling effective communication with them and their loved ones (Rosas et al., 2003). Our routine includes multidomain assessment, which encompasses assessments of global cognition, executive function, learning and memory, attention, language, depression, apathy and social cognitivism (Stoker et al., 2022). Citalopram and sertraline are examples of selective serotonin reuptake inhibitors (SSRIs) that can be used for depression, anxiety or to treat psychosis, while atypical antipsychotics may be utilized for psychotherapy. In cases of marked obsessive-compulsive disorder (OCD), sertraline may be effective and used as a treatment option. It is possible to use medications that can help stabilize the mood (lamotrigine and sodium valproate) or tricyclic antidepressants, mirtazapine and benzodiazepines as an adjunct (Anderson et al., 2011).

Patients who experience significant sleep disturbances can benefit from receiving sleep hygiene education, which involves avoiding morning napping, engaging in regular exercise, eating healthily and restricting caffeine consumption, tobacco use, alcohol intake or screen time before bed. Medications should be evaluated to identify those that may worsen sleep problems, and the advantages and disadvantages of each should also be considered. The use of modafinil during daytime to ease sleep can help restore lost focus and improve overall health. Despite its limited availability in the UK and lack of evidence, mirtazapine can be helpful in treating depression and irregular sleep patterns, while melatonin may help restore a normal circadian rhythm. Certain Z-drugs may be beneficial, and we've also utilized sodium oxybate in individuals with poor sleep refractory to other treatments (Stoker et al., 2022).

REFERENCES

Anderson, K., Craufurd, D., Edmondson, M. C., Goodman, N., Groves, M., van Duijn, E., van Kammen, D. P., & Goodman, L. (2011). An international survey-based algorithm for the pharmacologic treatment of obsessive-compulsive behaviors in Huntington's disease. *PLoS Currents*, *3*, RRN1261. https://doi.org/10.1371/currents.RRN1261

Andrew, S. E., Goldberg, Y. P., Kremer, B., Telenius, H., Theilmann, J., Adam, S., Starr, E., Squitieri, F., Lin, B., & Kalchman, M. A. (1993). The relationship between trinucleotide (CAG) repeat length and clinical features of Huntington's disease. *Nature Genetics*, *4*(4), 398–403. https://doi.org/10.1038/ng0893-398

Asmal, L., Flegar, S. J., Wang, J., Rummel-Kluge, C., Komossa, K., & Leucht, S. (2013). Quetiapine versus other atypical antipsychotics for schizophrenia. *The Cochrane Database of Systematic Reviews*, (11), CD006625. https://doi.org/10.1002/14651858.CD006625.pub3

Aylward, E. H., Sparks, B. F., Field, K. M., Yallapragada, V., Shpritz, B. D., Rosenblatt, A., Brandt, J., Gourley, L. M., Liang, K., Zhou, H., Margolis, R. L., & Ross, C. A. (2004). Onset and rate of striatal atrophy in preclinical Huntington disease. *Neurology*, *63*(1), 66–72. https://doi.org/10.1212/01.wnl.0000132965.14653.d1

Bachoud-Lévi, A. C., Ferreira, J., Massart, R., Youssov, K., Rosser, A., Busse, M., Craufurd, D., Reilmann, R., De Michele, G., Rae, D., Squitieri, F., Seppi, K., Perrine, C., Scherer-Gagou, C., Audrey, O., Verny, C., & Burgunder, J. M. (2019). International guidelines for the treatment of Huntington's disease. *Frontiers in Neurology*, *10*, 710. https://doi.org/10.3389/fneur.2019.00710

Bang, J., Spina, S., & Miller, B. L. (2015). Frontotemporal dementia. *Lancet (London, England)*, *386*(10004), 1672–1682. https://doi.org/10.1016/S0140-6736(15)00461-4.

Boxer, A. L., Knopman, D. S., Kaufer, D. I., Grossman, M., Onyike, C., Graf-Radford, N., Mendez, M., Kerwin, D., Lerner, A., Wu, C. K., Koestler, M., Shapira, J., Sullivan, K., Klepac, K., Lipowski, K., Ullah, J., Fields, S., Kramer, J. H., Merrilees, J., Neuhaus, J., … Miller, B. L. (2013). Memantine in patients with frontotemporal lobar degeneration: A multicentre, randomised, double-blind, placebo-controlled trial. *The Lancet Neurology*, *12*(2), 149–156. https://doi.org/10.1016/S1474-4422(12)70320-4

Burgunder, J. M., Guttman, M., Perlman, S., Goodman, N., van Kammen, D. P., & Goodman, L. (2011). An international survey-based algorithm for the pharmacologic treatment of chorea in Huntington's disease. *PLoS Currents*, *3*, RRN1260. https://doi.org/10.1371/currents.RRN1260

Burrell, J. R., Kiernan, M. C., Vucic, S., & Hodges, J. R. (2011). Motor neuron dysfunction in frontotemporal dementia. *Brain: A Journal of Neurology*, *134*(Pt 9), 2582–2594. https://doi.org/10.1093/brain/awr195

Capell, A., Liebscher, S., Fellerer, K., Brouwers, N., Willem, M., Lammich, S., Gijselinck, I., Bittner, T., Carlson, A. M., Sasse, F., Kunze, B., Steinmetz, H., Jansen, R., Dormann, D., Sleegers, K., Cruts, M., Herms, J., Van Broeckhoven, C., & Haass, C. (2011). Rescue of progranulin deficiency associated with frontotemporal lobar degeneration by alkalizing reagents and inhibition of vacuolar ATPase. *The Journal of Neuroscience: The Official Journal of the Society for Neuroscience*, *31*(5), 1885–1894. https://doi.org/10.1523/JNEUROSCI.5757-10.2011

Cenik, B., Sephton, C. F., Dewey, C. M., Xian, X., Wei, S., Yu, K., Niu, W., Coppola, G., Coughlin, S. E., Lee, S. E., Dries, D. R., Almeida, S., Geschwind, D. H., Gao, F. B., Miller, B. L., Farese, R. V. Jr, Posner, B. A., Yu, G., & Herz, J. (2011). Suberoylanilide hydroxamic acid (vorinostat) up-regulates progranulin transcription: Rational therapeutic approach to frontotemporal dementia. *The Journal of Biological Chemistry*, *286*(18), 16101–16108. https://doi.org/10.1074/jbc.M110.193433

Chow, T. W., Miller, B. L., Hayashi, V. N., & Geschwind, D. H. (1999). Inheritance of frontotemporal dementia. *Archives of Neurology*, *56*(7), 817–822. https://doi.org/10.1001/archneur.56.7.817

Dupuis, L., Gonzalez de Aguilar, J. L., Echaniz-Laguna, A., Eschbach, J., Rene, F., Oudart, H., Halter, B., Huze, C., Schaeffer, L., Bouillaud, F., & Loeffler, J. P. (2009). Muscle mitochondrial uncoupling dismantles neuromuscular junction and triggers distal degeneration of motor neurons. *PLoS ONE*, *4*(4), e5390. https://doi.org/10.1371/journal.pone.0005390

Eisen, A., Kim, S., & Pant, B. (1992). Amyotrophic lateral sclerosis (ALS): A phylogenetic disease of the corticomotoneuron? *Muscle & Nerve*, *15*(2), 219–224. https://doi.org/10.1002/mus.880150215

Gordon, P. H., Cheng, B., Katz, I. B., Pinto, M., Hays, A. P., Mitsumoto, H., & Rowland, L. P. (2006). The natural history of primary lateral sclerosis. *Neurology*, *66*(5), 647–653. https://doi.org/10.1212/01.wnl.0000200962.94777.71

Gorno-Tempini, M. L., Hillis, A. E., Weintraub, S., Kertesz, A., Mendez, M., Cappa, S. F., Ogar, J. M., Rohrer, J. D., Black, S., Boeve, B. F., Manes, F., Dronkers, N. F., Vandenberghe, R., Rascovsky, K., Patterson, K., Miller, B. L., Knopman, D. S., Hodges, J. R., Mesulam, M. M., & Grossman, M. (2011). Classification of primary progressive aphasia and its variants. *Neurology*, *76*(11), 1006–1014. https://doi.org/10.1212/WNL.0b013e31821103e6

Kiernan, M. C., Vucic, S., Cheah, B. C., Turner, M. R., Eisen, A., Hardiman, O., Burrell, J. R., & Zoing, M. C. (2011). Amyotrophic lateral sclerosis. *Lancet (London, England)*, *377*(9769), 942–955. https://doi.org/10.1016/S0140-6736(10)61156-7

Le Ber, I. (2013). Genetics of frontotemporal lobar degeneration: An up-date and diagnosis algorithm. *Revue Neurologique*, *169*(10), 811–819. https://doi.org/10.1016/j.neurol.2013.07.014.

Leigh, P. N., Abrahams, S., Al-Chalabi, A., Ampong, M. A., Goldstein, L. H., Johnson, J., Lyall, R., Moxham, J., Mustfa, N., Rio, A., Shaw, C., Willey, E., & King's MND Care and Research Team. (2003). The management of motor neurone disease. *Journal of Neurology, Neurosurgery, and Psychiatry*, *74*(Suppl 4), iv32–iv47. https://doi.org/10.1136/jnnp.74.suppl_4.iv32

Mackenzie, I. R., & Rademakers, R. (2008). The role of transactive response DNA-binding protein-43 in amyotrophic lateral sclerosis and frontotemporal dementia. *Current Opinion in Neurology*, *21*(6), 693–700. https://doi.org/10.1097/WCO.0b013e3283168d1d

Mesulam, M. M. (2001). Primary progressive aphasia. *Annals of Neurology*, *49*(4), 425–432.

Neary, D., Snowden, J. S., Gustafson, L., Passant, U., Stuss, D., Black, S., Freedman, M., Kertesz, A., Robert, P. H., Albert, M., Boone, K., Miller, B. L., Cummings, J., & Benson, D. F. (1998). Frontotemporal lobar degeneration: A consensus on clinical diagnostic criteria. *Neurology*, *51*(6), 1546–1554. https://doi.org/10.1212/wnl.51.6.1546

Olney, N. T., Spina, S., & Miller, B. L. (2017). Frontotemporal dementia. *Neurologic Clinics*, *35*(2), 339–374. https://doi.org/10.1016/j.ncl.2017.01.008

Rascovsky, K., Hodges, J. R., Knopman, D., Mendez, M. F., Kramer, J. H., Neuhaus, J., van Swieten, J. C., Seelaar, H., Dopper, E. G., Onyike, C. U., Hillis, A. E., Josephs, K. A., Boeve, B. F., Kertesz, A., Seeley, W. W., Rankin, K. P., Johnson, J. K., Gorno-Tempini, M. L., Rosen, H., Prioleau-Latham, C. E., … Miller, B. L. (2011). Sensitivity of revised diagnostic criteria for the behavioural variant of frontotemporal dementia. *Brain: A Journal of Neurology*, *134*(Pt 9), 2456–2477. https://doi.org/10.1093/brain/awr179

Rohrer, J. D., Guerreiro, R., Vandrovcova, J., Uphill, J., Reiman, D., Beck, J., Isaacs, A. M., Authier, A., Ferrari, R., Fox, N. C., Mackenzie, I. R., Warren, J. D., de Silva, R., Holton, J., Revesz, T., Hardy, J., Mead, S., & Rossor, M. N. (2009). The heritability and genetics of frontotemporal lobar degeneration. *Neurology*, *73*(18), 1451–1456. https://doi.org/10.1212/WNL.0b013e3181bf997a

Rosas, H. D., Koroshetz, W. J., Chen, Y. I., Skeuse, C., Vangel, M., Cudkowicz, M. E., Caplan, K., Marek, K., Seidman, L. J., Makris, N., Jenkins, B. G., & Goldstein, J. M. (2003). Evidence for more widespread cerebral pathology in early HD: An MRI-based morphometric analysis. *Neurology*, *60*(10), 1615–1620. https://doi.org/10.1212/01.wnl.0000065888.88988.6e

Rosen, H. J., Gorno-Tempini, M. L., Goldman, W. P., Perry, R. J., Schuff, N., Weiner, M., Feiwell, R., Kramer, J. H., & Miller, B. L. (2002). Patterns of brain atrophy in frontotemporal dementia and semantic dementia. *Neurology*, *58*(2), 198–208. https://doi.org/10.1212/wnl.58.2.198

Sanders, D. W., Kaufman, S. K., DeVos, S. L., Sharma, A. M., Mirbaha, H., Li, A., Barker, S. J., Foley, A. C., Thorpe, J. R., Serpell, L. C., Miller, T. M., Grinberg, L. T., Seeley, W. W., & Diamond, M. I. (2014). Distinct tau prion strains propagate in cells and mice and define different tauopathies. *Neuron*, *82*(6), 1271–1288. https://doi.org/10.1016/j.neuron.2014.04.047

Seeley, W. W., Crawford, R. K., Zhou, J., Miller, B. L., & Greicius, M. D. (2009). Neurodegenerative diseases target large-scale human brain networks. *Neuron*, *62*(1), 42–52. https://doi.org/10.1016/j.neuron.2009.03.024

Snowden, J. S., Gibbons, Z. C., Blackshaw, A., Doubleday, E., Thompson, J., Craufurd, D., Foster, J., Happé, F., & Neary, D. (2003). Social cognition in frontotemporal dementia and Huntington's disease. *Neuropsychologia*, *41*(6), 688–701. https://doi.org/10.1016/S0028-3932(02)00221-X

Stoker, T. B., Mason, S. L., Greenland, J. C., Holden, S. T., Santini, H., & Barker, R. A. (2022). Huntington's disease: Diagnosis and management. *Practical Neurology*, *22*(1), 32–41. https://doi.org/10.1136/practneurol-2021-003074

Talbot, K. (2009). Motor neuron disease: The bare essentials. *Practical Neurology*, *9*(5), 303–309. https://doi.org/10.1136/jnnp.2009.188151

Thompson, J. A., Cruickshank, T. M., Penailillo, L. E., Lee, J. W., Newton, R. U., Barker, R. A., & Ziman, M. R. (2013). The effects of multidisciplinary rehabilitation in patients with early-to-middle-stage Huntington's disease: A pilot study. *European Journal of Neurology, 20*(9), 1325–1329. https://doi.org/10.1111/ene.12053

Thompson, S. A., Patterson, K., & Hodges, J. R. (2003). Left/right asymmetry of atrophy in semantic dementia: Behavioral-cognitive implications. *Neurology, 61*(9), 1196–1203. https://doi.org/10.1212/01.wnl.0000091868.28557.b8

Turner, M. R., Kiernan, M. C., Leigh, P. N., & Talbot, K. (2009). Biomarkers in amyotrophic lateral sclerosis. *The Lancet. Neurology, 8*(1), 94–109. https://doi.org/10.1016/S1474-4422(08)70293-X

Vucic, S., & Kiernan, M. C. (2007). Abnormalities in cortical and peripheral excitability in flail arm variant amyotrophic lateral sclerosis. *Journal of Neurology, Neurosurgery, and Psychiatry, 78*(8), 849–852. https://doi.org/10.1136/jnnp.2006.105056

Walker, F. O. (2007). Huntington's disease. *Lancet (London, England), 369*(9557), 218–228. https://doi.org/10.1016/S0140-6736(07)60111-1

Zhang, Y., Schuff, N., Du, A. T., Rosen, H. J., Kramer, J. H., Gorno-Tempini, M. L., Miller, B. L., & Weiner, M. W. (2009). White matter damage in frontotemporal dementia and Alzheimer's disease measured by diffusion MRI. *Brain: A Journal of Neurology, 132*(Pt 9), 2579–2592. https://doi.org/10.1093/brain/awp071

Index

Note: Page numbers in **bold** and *italics* refer to tables and figures, respectively.

B

C

T

U

V

W

X

Z